INTRODUCTION TO
Food Processing

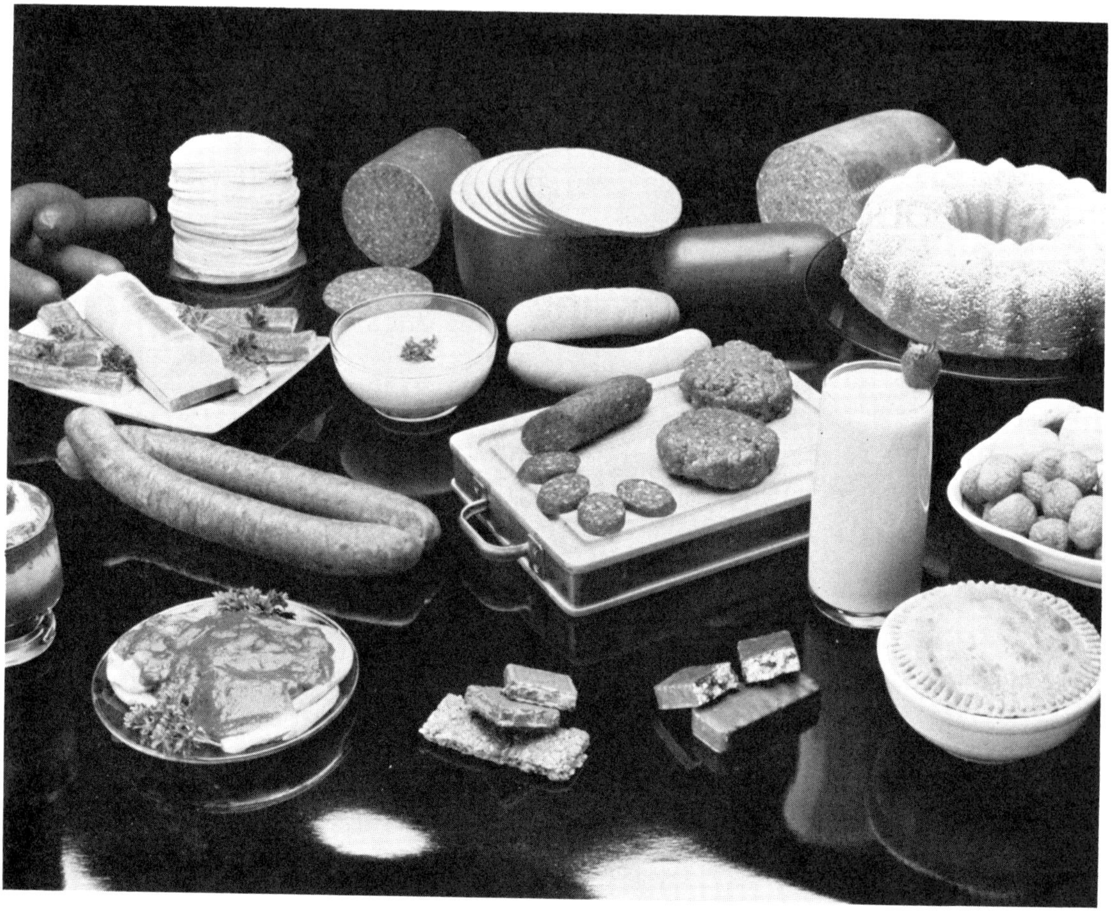

INTRODUCTION TO
Food Processing

P. Jelen
PROFESSOR OF FOOD SCIENCE
UNIVERSITY OF ALBERTA
EDMONTON, ALBERTA, CANADA

RESTON PUBLISHING COMPANY, INC.
A PRENTICE-HALL COMPANY
RESTON, VIRGINIA

Library of Congress Cataloging in Publication Data

Jelen, Pavel.
  Introduction to food processing.

  1. Food industry and trade.   I. Title.
TP370.J45   1985   664   84-27704
ISBN 0-8359-3194-3

© 1985 by Reston Publishing Company, Inc.
A Prentice-Hall Company
Reston, Virginia 22090

All rights reserved. No part of this book may be reproduced in any way,
or by any means, without permission in writing from the publisher.

1 3 5 7 9 10 8 6 4 2

PRINTED IN THE UNITED STATES OF AMERICA

TO   My Parents, who taught me to appreciate food as one of the gifts of life;

AND

TO   Sylva, whose patience, support and understanding was my main source of encouragement.

Contents

Preface xi

1 What Is Food? 1
Food in a Modern Society 1
Agricultural Production of Raw Food Materials 2
Food Composition 4
Macronutrients and Micronutrients 7
Nutritional Requirements of a Human Body 13
Nutritive Value of Processed Foods 19
Industrial Food Processing and Health Foods 24
Control Questions 26
Suggested Reference Books 27

2 Microorganisms in Foods 29
Forms of Microbial Life 29
Microbial Growth and Food Spoilage 31
Water Activity 34
Food Poisoning or Food-Borne Illness 38
Detection and Enumeration of Microorganisms 40
Food Fermentations 42
Role of Microorganisms in Human Nutrition 44
Control Questions 45
Suggested Reference Books 47

3 The Food Industry   49

Food Manufacturing and Preservation   49
Food Processing as an Applied Science   51
Industrial Plant Organization   53
Product Development and Quality Control   59
Government Regulations   62
Material and Energy Balances   65
Solubility and Concentration in Liquid Foods   66
Control Questions   68
Suggested Reference Books   70

4 Processing of Fruits and Vegetables   71

Characteristics of Fresh Fruits and Vegetables   71
Photosynthesis and Respiration   73
Industrial Harvesting and Ripening   73
Fruit and Vegetable Storage   75
Preparative Processes   77
Blanching   78
Fruit and Vegetable Products   81
Jams and Jellies   83
Manufacture and Crystallization of Sugar   86
Nutritional Importance of Fruit and Vegetable Products   88
Control Questions   90
Suggested Reference Books   91

5 Cereal Grains and Oilseeds   93

Raw Materials   93
Cereal Grains: Kernel Structure and Milling   94
Baking Bread   99
Pastry, Cookies and Pasta Products   105
Milling and Processing of Other Cereals   106
Malting   107
Oilseeds: Composition and Extraction of Vegetable Oils   109
Margarine and Other Vegetable Oil Products   114
Oil Emulsions   117
Nutritional Aspects of Cereal and Oilseed Products   119
Control Questions   122
Suggested Reference Books   122

6 Milk and Dairy Products   125

Milk Production   125
Milk Composition   127

Fluid Milk Processing 127
Cultured Dairy Products 133
Cheese 137
Butter 141
Ice Cream 145
Concentrated, Sterilized and Dried Products 146
Nutritive Value of Milk and Dairy Foods 147
Control Questions 149
Suggested Reference Books 151

7 Processing of Meat, Poultry, and Fish 153

Sources and Production 153
Meat Composition 154
Living Muscle and Rigor Mortis 155
Fresh Meat 158
Curing 162
Processed Meat Products 166
Poultry Processing 169
Fish and Seafood 170
Nutritive Value of Red Meats, Poultry and Fish 172
Control Questions 173
Suggested Reference Books 174

8 Alcoholic and Non-Alcoholic Beverages 175

Industrial Beverage Products 175
Alcoholic Fermentation 176
Beer and the Brewing Process 178
Wine and Liquor Products 185
Coffee and Tea 189
Soft Drinks 192
Nutritional and Medical Implications of Industrial Beverages 195
Control Questions 197
Suggested Reference Books 198

9 Food Additives and Food Ingredients 199

What are Food Additives? 199
Functional Properties 200
Food Ingredients 202
Food Additives 205
Procedures for Testing Food Additives 208
Chemical Preservatives 210

Agricultural Chemicals and Other Food Contaminants 212
Labeling Requirements for Additives and Ingredients 213
Effects of Food Additives in Human Nutrition 214
Control Questions 218
Suggested Reference Books 218

10 Processes for Food Preservation 221

Causes and Prevention of Food Spoilage 221
Food Dehydration 223
Food Freezing 230
Heat Sterilization and Canning 236
Other Food Preservation Processes 241
Food Irradiation 243
Nutritional Significance of Long-Term Food Preservation 246
Control Questions 246
Suggested Reference Books 247

11 Food Packaging Technology 249

Purpose of Food Packaging 249
Properties and Uses of Food Packaging Materials 251
Laminated Packaging Materials 259
Effects of Packaging on Food Stability 262
Modified Atmosphere Packaging 263
Role of Packaging in the Food Distribution Chain 264
Control Questions 265
Suggested Reference Books 266

Appendix I:
Procedures for Mass and Energy Balance Calculations 267
Appendix II:
Selected Literature Topics 277
Appendix III:
List of Commonly Used Food Additives and
Explanation of Their Functions 301
Index 307

Preface

The proliferation of new food products on supermarket shelves is one of the indicators of technological advancement and increased living standards of a modern society. However, unlike new developments in other areas of consumer interest, the innovations of industrial food technology are often received with suspicion or even hostility. There exists a communication gap between trained food technologists who translate rapid advances in scientific knowledge into benefits for mankind, and the benefactor of the efforts, the consuming public. Sometimes, food scientists and students of food science are fascinated with the most sophisticated laboratory techniques and industrial methodology for space-age foods, while knowledge of the traditional food processing technology is being taken for granted. On the other hand, consumers need to understand at least the basic manufacturing principles for the most common food products, to be able to appreciate the benefits of industrially processed foods for affordable, yet balanced nutrition enjoyed by the modern society today.

This book is an attempt to bridge the gap between the serious student of food technology and the interested consumer. The writing is based on ten years of experience in teaching an introductory food processing course to university students—both majors and non-majors in Food Science—as well as teaching an extension food processing course for the general public. Whether my attempt to write for both groups of readers was a futile attempt of the impossible awaits the judgement of the reader. However, the deliberate selection of emphasis meant that complicated technological topics needed to be explained in simplified form or sometimes avoided altogether. This was considered a necessary trade-off between thoroughness of coverage required for a university text, and clarity of expression understandable without complex scientific background. The objective of my writing was to provide a comprehensive, yet simple description of the "how" and "why" of the manufacture of most commonly consumed foods. My aim was to introduce concepts of a large number of contemporary food processing practices without much technical detail. It is expected that the inquisitive reader—or the concerned teacher—will search for additional information to complement the text. Selected lists of books covering topics of each chapter are provided for this purpose and not as reference material for specific data. The lists do not contain all books available on a given topic and indicate a personal preference rather than a selection of most appropriate sources. Product definitions were generalized from the Canadian Food and Drug Act and Regulations to minimize discrepancies among food laws of various countries.

Numerous individuals deserve a deep expression of gratitude for their input without which this book could not have been completed. Much illustrative material and other suitable information has been provided by industrial companies or individuals from universities and various research establishments; their kind contributions are acknowledged throughout the text. Several colleagues and friends checked early drafts of the individual chapters and suggested numerous improvements. Special thanks in this regard are due to Professors Addis, Breene and Busta from the University of Minnesota; Bhowmik, Morr and Thomas at Clemson University; Ernstrom at Utah State University; Bligh at Nova Scotia Technical University; Basu and Ooraikul at the University of Alberta; Drs. G. Timbers of Eng. and Stat. Research Institute, Ottawa, and E. Jackson of Res. Station Kentville, Nova Scotia, both of Agriculture Canada; and Messrs T. Schmidt of J. Labatt's Co. Creston, B.C., and T. Warwaruk of The Pop Shoppe Co., Edmonton. Helpful reviews of the completed manuscript by Professors J. Maga from the Colorado State University and S. Biede from Louisiana State University resulted in further improvements in some areas. Several "generations" of students from the Industrial Food Processing course at the University of Alberta provided valuable feedback and many suggestions concerning usefullness and readability of the book as a study manual. The patience, skill and efficiency of Candi Dubetz was of great help in the preparation of early drafts, while the thoroughness of both Candi and Donna Bornhuse was much appreciated in completion of the manuscript. Finally, thanks are expressed to the many colleagues and friends from industry in Canada and throughout the world who accomodated my numerous technical visits of their food processing facilities in the past several years; without their understanding and assistance this book would have been much less authentic.

P. Jelen

INTRODUCTION TO
Food Processing

1

What Is Food

1.1 FOOD IN A MODERN SOCIETY

Food is one of the three essential ingredients of human life. Together with water and oxygen, food provides the human body with organic and inorganic substances necessary to sustain its biological viability. The lack of food leads to hunger, to decreased functionality of the complicated biochemical systems of the body, and to nutritional diseases like rickets, pellagra or scurvy. Only in extreme cases will total lack of food result in cessation of biological functions and death. A human body can survive about a month without food, but only a few days without water, and a few minutes without air.

In the not so distant past, provision of food was the principal preoccupation of mankind. In the majority of the world's countries today, farming is still the most important activity. As the affluence of a society increases, the emphasis on food as an essential agricultural or industrial product becomes less significant. In North America, the availability of incredibly varied, year-round and affordable food supply has been taken for granted. Although hunger still exists, it is no longer associated with the inability to produce enough food due to natural disasters or overpopulation; rather, it is a social disease. The main concerns of the general population relating to food are no longer the magnitude of production, the protection from agricultural diseases and the rush to complete a harvest. Nowadays, we are much more concerned with the adequacy of nutrient supply. We have the luxury of selecting specific food items to receive a balanced diet; we become obese, a sure sign of abundance of basic nutrients; and we can afford to criticize the food processors for their efficiency and ingenuity in transforming the oversupply of agricultural produce into shelf-stable, convenient, colorful and flavorful industrial food products.

The general population in North America knows relatively little about the technology of food production, food manufacturing and food preservation. Ignorance breeds skepticism; the fear of the unknown has been one of the foundations of myths and fairy-tales. In a modern urbanized society, a general knowledge of agricultural practices and food manufacturing techniques is insufficient, leading to feelings of mistrust between the consumer and the food processing industry. This text will attempt to provide some of the simplest "*why's*" and "*how's*" of modern food processing, thus portraying foods as one of the essentials of life, to be enjoyed rather than condemned.

1.2 AGRICULTURAL PRODUCTION OF RAW FOOD MATERIALS

Most of the materials used as human food are derived from agricultural products. The two major categories of agricultural food materials are *plant products* and *animal products*. The concepts of breeding, production, harvesting, post-harvest storage, and material handling on the farm are traditionally associated with biological and agricultural sciences such as plant science, animal science, genetics, agricultural engineering, horticulture, etc. However, some of the concepts and the terminology used should be understood by the food processor and the consumer. Table 1-1 lists some of the commonly used agricultural terms that are of importance to food processors.

Factors associated with farm production such as yield, insect resistance, compatibility with different climates and soils, differences in composition, effectiveness in utilization of fertilizers or feeds, monetary return upon marketing, and many other factors will determine what type of commodity a farmer will produce. However, certain varieties of plants and breeds of animals are more suitable as fresh food for direct consumption without processing, while others are specifically developed by genetic breeding for industrial processing. Thus, certain varieties of potato (e.g. Norland or Pontiac) are suitable as a table potato due to their cooking characteristics, but poor for processing into French fries when high content of dry matter is required. Jersey or Guernsey breeds of dairy cattle produce milk with much higher fat content than Holsteins or Brown Swiss. In contrast, the latter are either better yielding or easily adaptable to certain terrain. If a bread is baked from wheat flours suitable for pasta products the result is an inferior product, as specific wheat varieties are suitable for specific industrial

Table 1-1 Explanation of Common Agricultural Terms

Variety (Cultivar) — Differentiation of plants of a particular species, based on variations in the genetic makeup. As a result, different varieties of the same species (e.g. tomato) may have different properties.

Breed — Similar differentiation used to describe genetic variations for animals of the same species.

Maturity — An optimum stage in the development cycle of a plant or an animal, relative to physiological or industrial aspects.

Climate — Interrelationships of conditions necessary for agricultural production including availability of moisture, amount of heat or sunlight, prevailing air movements, length of growing season.

Yield — Amount of useful agricultural material produced per one production unit (acre of land, one cow, unit of feed, etc.).

Breeding — Deliberate manipulation of the genetic makeup of a given plant or animal for creation of new varieties or breeds with improved characteristics (resistance, yield, maturity, etc.).

Harvest — Gathering of agricultural produce at its optimum maturity (physiological or industrial).

uses. The food processor must select judiciously what raw materials to use; buying just any kind of potato for a French fry processing factory could result in a poor quality product or uneconomic operation.

Harvesting, in the broadest sense of the word, refers to gathering of raw agricultural products (fruits, vegetables, wheat, cattle, chicken) by farmers. In many cases, harvesting for industrial food processors will be done at a different stage of physiological maturity than harvesting for fresh product markets. The food processor can influence the harvesting operation by requesting his raw material to be at a specific maturity, specific temperature, or specific hygienic standard. Collaboration between farmers and industrial processors is essential in this regard.

Some varieties of fruits and vegetables (strawberries, tomatoes) have been improved by breeding to be specifically suitable for mechanical harvesting (see Chapter 4). This is important in view of the ever-rising cost of labor. Without the "square tomatoes" as they are sometimes called, the price of ketchup would soon become prohibitively high. Similar economic arguments can be made in support of other agricultural processes, for example, the breeding of chickens adapted to the "assembly-line" style of production of poultry and eggs. The genetic make-up of these breeds is such that they can survive the stressful situation in which they have to live, yet produce acceptable and cheap meat and eggs.

The raw food materials produced by the farmer can rarely be used without some modification. There are many inedible parts of the plant or animal material that must be discarded even in the farm kitchen if the farmer uses his production as his source of food. Many farm commodities serve as sources of industrial products that must be processed further. Thus, flour or white granulated sugar are not particularly suitable foods as such. The fundamental function of food processing is to convert the maximum amounts of nutrient components of the agricultural raw materials into forms suitable for human consumption as food.

1.3 FOOD COMPOSITION

From the chemist's viewpoint, all organic or inorganic matter, including all agricultural materials and foods made from them, are composed of chemicals. As will be shown in Chapter 4, the principal building blocks of all plant materials are simple chemical compounds such as oxygen or carbon dioxide from the air; phosphorus, nitrogen and potassium from the soil or the fertilizer used to improve the soil productivity; water; and other chemical elements or their simple combinations. These and similar chemical building blocks are used by all living organisms (microorganisms, plants, animals, or humans) to synthesize much more complex chemicals like proteins, fats, sugars or vitamins which comprise the final chemical make-up of our foods. The bulk of food components are complex organic compounds built from only a few principal elements—carbon, oxygen, hydrogen, and nitrogen—bound in almost infinite variety of chemical combinations. The various chemical configurations result in the fundamentally different physicochemical and nutritional properties of proteins, fats, or carbohydrates. Several other elements (especially calcium, phosphorus, iron, or sulphur) may be also involved in the complex chemical structures of these organic compounds, while other elements (sodium, potassium, chlorine, zinc, magnesium and many more) are present in the food material as free ions, salts or other simple inorganic compounds, often dissolved in the main food constituent—water.

There are only five major categories of distinctly different chemical compounds that all foods are composed of: proteins, carbohydrates (CHO), fats, minerals (ash), and water. When a food is broken down into these five kinds of components, without differentiating as to what specific molecules of various proteins, fats, carbohydrates or minerals are present, we speak of *proximate* analysis or proximate composition. Although from a nutritional standpoint vitamins are classified as a separate nutrient group, in a proximate analysis of foods they will be included with some of the principal proximate components, primarily because of their chemical similarity and/or their presence in very minute quantities. In rare instances there are other chemical food components that will not be determined by the customary proximate analysis procedures; the most notable examples are alcohol and CO_2 in beverages (Chapter 8). In most cases, a proximate analysis will account for 100 percent of the material's composition.

Proximate composition is always given in weight percentages. It can include the water, so that the weight of all the five components present will add up to 100 percent. This is referred to as "wet basis," or "as is." When the water has been removed, the remaining four components are sometimes called *dry matter*, or *total solids* (T.S.). Their percentage proportion in the dry matter is often expressed as proximate composition —*dry basis* (d.b.). In some cases, dry matter composition may be broken down further to percentage of fat in the total solids, and the percentage of *non-fat solids*, comprising the total of CHO, protein, and minerals.

Each component group of the proximate composition may include chemically different molecules showing the same overall structural pattern. Thus, the total protein content of milk is about 3.2 percent, but a more detailed investigation would reveal at least three major protein fractions differing in molecular structure and some physico-chemical properties (casein, α-lactalbumin, β-lactoglobulin), as well as several other minor dairy proteins. However, all these chemically different molecules have the same overall chain-like amino acid structure (see Section 1.5), which differentiates them from other classes of organic compounds.

Similarly, butterfat and sunflower oil are both composed of various fatty acids bound to molecules of glycerol (see Section 5.7). This makes the fats as a group different from sugars, but because the kinds and amounts of the individual fatty acids are different in different fats, butterfat has some distinctly different properties from sunflower oil or lard. The diverse group of carbohydrates includes various simple sugars, starches, and indigestible materials like cellulose or lignin which, from the nutritional standpoint, may be more properly classified as dietary fiber. Many simple inorganic chemicals found in foods are included in the group of minerals; from the standpoint of proximate analysis we often speak of ash since the laboratory technique used for minerals' analysis consists of evaporating all water and burning off all organic materials. The remaining ash contains all the minerals originally present in the food, whether dissolved as simple ions or present in other forms.

All proximate analyses are carried out by simple chemical tests based on *gravimetry*, or weight differences. Representative food samples of known weight are subjected to a simple chemical extraction, or elimination of all competing substances, and the results are expressed as percentage of the original weight. Table 1–2 gives a brief description

Table 1-2 Basic Methods for Proximate Analysis of Foods

Component	GENERAL METHOD Common Name	Principle[a]	Reference Example[b]
Protein	Kjeldahl	Digest a food sample by sulphuric acid; determine nitrogen (N_2) content in the digest; convert the N_2 content into protein content by calculation.	Method 47.021
Fat	Soxhlet	Extract all lipid from a food sample by organic solvent (e.g. petroleum ether); determine weight of extracted fat	Method 27.006
Water	Vacuum Oven	Evaporate water from a food sample to dryness; determine weight difference.	Method 7.003
Minerals (ash)	Muffle Furnace	Burn a food sample to CO_2, H_2O and ash; determine weight of ash.	Method 31.012
Carbohydrate	—	Determine as difference between 100% and total of all other proximate components.	

[a] Final results expressed as percentage of original weight of a food sample.
[b] *Official Methods of Analysis*, Washington, D.C.: The Association of Official Analytical Chemists, 1984, 14th edition.

of the proximate analysis techniques used in routine quality control work.

The knowledge of proximate composition of raw materials as well as of processed foods is important for both the food processors and the consumers. Thus, the dairy processor must know how much fat there is in the milk that he is going to process into butter. By knowing how much milk he has available, he can then design, order, and operate his processing line. He can also account for and minimize his losses by comparing how much butter he produced, and how much he should have produced from the butterfat that he had available.

Similarly, processors of grains, oilseeds, fruits and vegetables, meats and other agricultural products must know the proximate composition of their raw materials, especially when there are large differences related to production techniques, varieties or breeds concerned, and other factors. Furthermore, the proximate composition, especially the moisture content (percentage of H_2O) of various food materials, is critically important for their storage stability and the need for special storage precautions. In general, the only storage-stable raw materials are those with low H_2O content (typically below 10 percent) such as grains, pulses and oilseeds. Orientation values of proximate composition for major raw agricultural products of importance in food processing are listed in Table 1–3.

Proximate composition of finished food products is also of major interest to both processors and consumers. Many processed foods must conform to legal definitions which are usually given in terms of proximate composition. Using once again butter as an example, it must legally contain at least 80 percent butterfat; in some countries it also must contain not more than 16 percent water. Similarly, processed meat products, cottage cheese, ice cream, jams and jellies, and many other products are legally defined in terms of maximum allowable water, minimum required protein, or specific amounts of fat. Table 1–4 gives orientation data for a few representative processed foods.

In some European countries, proximate composition data must be displayed on most food products sold to the public. In view of the current consumer concerns relating to nutrition, a similar requirement for such simple nutritional information should be mandatory in the United States and Canada. The existing or proposed nutrition labelling programs in both countries are less satisfactory since they require data for many specific nutrients in addition to the main proximate components. Furthermore, labelling is required only on those foods for which a nutritional claim is made. The requirement that proximate analysis data be displayed on all processed foods would provide much more meaningful nutrition information than the selectively applied voluntary program of today.

1.4 MACRONUTRIENTS AND MICRONUTRIENTS

Quantitatively, foods consist principally of three macronutrients—proteins, fats (lipids) and carbohydrates. Together with water, which could be also considered a macronutrient, these components may constitute more than 99 percent of the foods (Tables 1–3 and 1–4). The three principal macronutrients supply the human body with energy needed for its physical and mental activity, and, in the case of protein, also with

Table 1-3 Proximate Composition of Some Raw Food Materials (Orientation Values)[1]

	H_2O	PROTEIN	FAT	CHO (INCLUDING FIBER)	ASH
	\% TOTAL WEIGHT					

a) Plant Origin
Fruits and Vegetables

	H_2O	PROTEIN	FAT	CHO	ASH
Potato	80.0	2.0	0.1	17.0	0.9
Apple	84.0	0.2	0.6	15.0	0.3
Tomato	93.0	1.0	0.2	5.0	0.5
Lettuce	95.0	1.0	0.2	3.0	1.0
Orange	86.0	1.0	0.2	12.0	0.6
Green Peas	78.0	6.0	0.4	15.0	1.0

Grains, Pulses & Oilseeds (Dry)

Rice	12.0	7.0	0.4	80.0	0.5
Wheat	13.0	14.0	2.2	69.0	1.7
Soybean	8.0	40.0	18.0	30.0	4.0
Rapeseed	7.0	18.0	45.0	26.0	4.0
Sunflower	4.0	23.0	50.0	20.0	3.0
Peanuts	5.0	26.0	48.0	19.0	2.0

b) Animal Origin
Meat and Poultry (Average Values)

Beef	60.0	18.0	21.0	trace	1.0
Pork	37.0	10.0	52.0	trace	0.5
Chicken	75.0	19.0	5.0	trace	0.8
Turkey	64.0	20.0	15.0	trace	1.0
Duck	62.0	21.0	16.0	trace	1.0

Fish - Seafood

Cod	81.0	18.0	0.3	trace	1.0
Fatty Fish (Eel)	65.0	16.0	18.0	trace	0.2
Oyster	84.0	9.0	2.0	3.0	2.0
Shrimp	78.0	18.0	1.0	1.5	1.5

Eggs - Whole

(Without Shell)	73.0	13.0	12.0	1.0	1.0
White (58% Total Weight)	88.0	11.0	trace	0.8	0.7
Yolk (31% Total Weight)	52.0	16.0	30.0	0.5	1.5

Milk

Cow's	87.0	3.5	3.5–5.0	5.0	0.7
Goat's	87.0	3.2	4.0	4.5	0.7
Human	85.0	1.0	4.0	9.5	0.2

[1] Data from various sources.

Table 1-4 Proximate Composition of Some Processed Foods (Orientation Values)[1]

	H_2O	PROTEIN	FAT	CHO (INCLUDING FIBER)	ASH
	\% TOTAL WEIGHT				
White Bread	36.0	9.0	3.0	50.0	2.0
Cheddar Cheese	37.0	25.0	32.0	2.0	4.0
Bologna	57.0	13.0	23.0	4.0	3.0
Strawberry Jam	29.0	0.5	0.1	70.0	0.3
Frozen Pizza With Cheese	50.0	12.0	8.0	27.0	3.0
Beer	92.0	0.3	0.0	4.0	0.2
Breakfast Cereals (Cornflakes, Unsweetened, Enriched)	4.0	8.0	0.5	85.0	2.5
Ice Cream	63.0	4.0	11.0	21.0	1.0

[1] Source: C.F. Adams, *Nutritive value of American Foods*, Washington, D.C.: (Agric. Handbook No. 456, USDA).

principal building blocks (amino acids) needed for development, maintenance and repair of its tissues. The traditional goal of food processing has been the utilization and preservation of the macronutrients supplied by the farmer in his produce.

The micronutrient group of compounds contains primarily vitamins and minerals. Contrary to macronutrients, the micronutrients are present in raw agricultural materials in minute quantities. Since minerals as a group are chemically stable, they too, like macronutrients, will be preserved in many processed foods. Some losses of minerals, which are soluble in water, may occur in those processes where the food material comes into prolonged contact with water. While most of the nutritionally important minerals are present in foods in simple elemental forms, the vitamins are complex organic compounds which cannot be synthesized by the human body but which are needed for its many biological functions. The complex chemical structures and consequently the biological activity of the vitamins may be sensitive to various environmental conditions encountered in food processing, storage, and home food preparation. Certain vitamins may be destroyed by light, high temperature, or specific oxidative reactions, or removed in processing with water (H_2O soluble vitamins) or in various fat separation processes (the fat-soluble vitamins). Table 1-5 gives a list of the fourteen currently recognized vitamins, some of their principal food sources, and some of the factors which influence their stability.

Vitamins, as well as some minerals, participate in numerous bio-

Table 1-5 Vitamins - Some Properties and Nutritional Requirements

VITAMINS					Approximate Dietary Requirements for Adults[c] mg/Day	
Preferred Name	Alternate Name[a]	Solubility[b] in Water (W) or Fat (F)	Principal Agents of Destruction	Best Food Sources		Notes
Vitamin A	Carotene, retinol	F	Light, oxygen	Milkfat, eggs, yellow and green vegetables	1.0	Toxic in large amounts
Thiamin	B-1	W	Dry heat	Meats, organs, nuts, legumes, whole cereals	1.2	Toasting bread may cause significant losses
Riboflavin	B-2	W	Light	Milk and dairy products, meats, green vegetables	1.5	Destroyed in milk if not protected from light
Pantothenic acid	B-3	W	Acid environment	Many foods (present in cells of all living organisms)	4-7	No recommended dietary allowance as intake exceeds natural needs
Niacin	B-5, Nicotinamide	W	—	Protein foods, especially meats and oilseeds	18.0	Most stable vitamin
Vitamin B-6	Pyridoxine	W	Light, heat	Meats, nuts, seeds, avocado, banana	2.2	May be lost in milling as fortification of white flour with this vitamin is not mandatory
Biotin	—	W	Oxygen	Many foods	?	Synthesized by intestinal bacteria so supply adequate; no recommended allowance

Choline	—	W		Eggs, meats, whole grains, many other foods	?	Importance as vitamin for humans is questioned, as it can be synthesized by the body
Folacin	Folic acid	W	Heating, especially in acid environment	Green leafy vegetables, oranges, legumes	0.2	Destroyed in vegetables upon storage
Vitamin B-12	Cyanocobalamine	W	Light, heat	Animal foods, esp. seafood, organs and meat	0.03	Synthesized only by microorganisms and fungi
Vitamin C	Ascorbic acid	W	Oxygen, enzymatic reactions, heat	Citrus fruits, green vegetables, other fruits and vegetables	60.0	Short heating may increase retention as enzymes are destroyed
Vitamin D	Cholecalciferol	F	UV light	Fortified milk, fatty fish, liver	0.01	Exposure to sunlight triggers synthesis in the human body
Vitamin E	α-Tocopherol	F	Oxygen, UV light	Cooking oils, margarine, nuts, oilseeds	10.0	Many unproven health-related claims for large dose intake
Vitamin K	—	F	Light	Green vegetables, soybeans, eggs	0.01	Synthesis by intestinal bacteria supplies most human needs

[a] Although often used as synonyms, most of these terms are inaccurate, obsolete, or identify only one of several compounds contributing to a particular vitamin complex.
[b] Water solubility indicates possible leaching losses in food processes where water is used; fat solubility may result in losses due to mechanical separation of fats.
[c] Adapted by author from recommended dietary allowances for various age and sex groups (issued by National Academy of Sciences, Food and Nutrition Board, Washington, D.C., 1980), and from other sources.

chemical reactions which control proper conduct of our physical and mental activities. For example, vitamin D facilitates the absorption of dietary calcium and phosphorus in the intestine, and regulates the blood levels of calcium needed for essential biochemical processes such as the proper functioning of the heart. Pantothenic acid is needed as a part of a complex biological catalyst (enzyme) which controls energy metabolism and synthesis of several essential substances including cholesterol, fatty acids and hormones. Vitamin B-6 is involved in regeneration and proper functioning of nerve tissues.

In these and many other biochemical processes the vitamins are used up; however, the amounts needed to be supplied daily are very small (Table 1-5) compared to the macronutrient requirements. Most of the dietary requirements for vitamins are amply provided by the many fresh and processed food materials that are a common part of the western-style diet. The widespread belief that processed foods are nutritionally inadequate because they lack vitamins is symptomatic of the poor understanding, prevalent among the general population, of the mechanism of vitamin functions and associated levels of dietary requirements, since "good nutrition" is often equated with "plentiful supply of vitamins." The ever growing use of vitamin supplement pills results in staggering (over $1.2 billion annually in the U.S. alone) and largely unnecessary expenditures, since the vitamins consumed in excess of biological needs are simply not utilized. The water-soluble vitamins may be excreted in urine (it is sometimes said that the U.S. sewage system is one of the richest vitamin sources!); however, the fat-soluble vitamins tend to accumulate in the fatty tissue deposits of our bodies and in excessive amounts may show toxic effects. Overdoses of various vitamins have been identified as causes of several health disorders such as kidney stones (vitamins C or D), internal bleeding (vitamin C), or even death (vitamin A).

As noted in Table 1-5, some of the vitamins are synthesized by our intestinal bacteria, while others are plentiful in many different food materials. Surveys conducted by the U.S. Department of Agriculture in 1977 and by Health and Welfare Canada from 1970 to 1973 documented that the average North American diet adequately supplies virtually all micronutrient needs of the various age and sex groups.

Modern food processing techniques are striving to minimize the destruction of the micronutrient content while still accomplishing the principal objective of maximizing the utilization of the macronutrients. In addition, as our knowledge of micronutrient requirements of the human body increases, certain foods are being fortified with specific minerals or vitamins to replace those that may be inevitably lost in the processing.

1.5 NUTRITIONAL REQUIREMENTS OF A HUMAN BODY

Unlike plants, the human body cannot synthesize complex organic substances from simple inorganic compounds and water coming from the soil or the air. All materials needed for proper functioning of the human body must be supplied orally, in the food we eat and the beverage we drink. The individual components of our foods are broken down during the various digestive processes in the stomach, and new, often much more complex compounds, are synthesized to be used in numerous metabolic reactions proceeding simultaneously in the body. Some substances, such as the vitamins, are absorbed from the food unchanged.

Food supplies nutrients that are required for three principal functions—to provide energy, to facilitate development and maintenance of bones, muscles and organs, and to enter into many biochemical reactions controlling our behavior or the proper functioning of organs. Contemporary nutrition science has identified about fifty substances that our body requires from food because it cannot synthesize them; these substances (among them several amino acids from proteins, one or two fatty acids from fats, as well as vitamins and some minerals) are termed *essential nutrients*. Because of the "buffering effect" that the storage capacity of the body has in terms of nutrient requirements, it is not necessary—and virtually impossible—to supply all these essential nutrients in all foods, all the time. The balanced diet available in affluent societies is an adequate source of all the essential nutrients, so that no medical supplementation is necessary. No widespread occurrence of the traditional nutritional diseases, listed in Table 1–6 together with their causes, is prevalent in the western world with the possible exception of osteoporosis and osteomalacia. In the past, and in many underdeveloped countries of the world even today, the supply of enough energy to sustain life is the principal nutritional requirement. On the contrary, a major contemporary medical problem of industrialized countries where food is plentiful is obesity, a direct result of consuming more food energy than the body needs for its various functions. The excess energy is not excreted, as water can be if one drinks too much, but is deposited by the body as fat.

The principal sources of food energy are carbohydrates and fats. The energy value of all foods can be determined fairly accurately from their proximate composition. The conversion factors for amounts of energy provided by the fats, carbohydrates and proteins (if used as an energy source which is less common) are listed in Table 1–7. Water and minerals do not contribute any energy. Alcohol, which is also listed in

Table 1–6 Major Nutritional Deficiency Diseases and Their Causes

DISEASE	MAJOR SYMPTOMS	CAUSE
Scurvy	Hemorrhagic skin, gangrene of gums and loss of teeth, pains in joints	Lack of vitamin C
Rickets	Weak bones, leg deformities ("bow legs"), distorted ribs (children)	Lack of vitamin D and calcium
Osteoporosis and Osteomalacia	Weakening of bone; pathological changes of bone, complications with healing fractured bones (adults)	Chronic lack of calcium, often due to lack of vitamin D
Pellagra	Skin rash, dark and rough skin, bloody diarrhea, mental disorder	Lack of niacin
Beri beri	Leg cramps, atrophy of leg muscles, paralysis, nervous and cardiac disturbances	Lack of thiamin
Anemia	Reduced level of hemoglobin in blood, paleness of skin, shortness of breath	Lack of iron
Goiter	Enlargement of thyroid gland (neck), birth defects in newborns	Lack of iodine
Kwashiorkor	Skin rash, orange color of hair, retarded growth (children)	Lack of protein
Marasmus	Atrophy of muscles, old age look (children), "skeleton" appearance	Lack of all nutrients

the table, is not a common food component, but a fairly important source of food energy. Table 1–7 also gives energy values of some common foods, while the energy requirements of some typical human activities are given in Table 1–8.

Protein and some minerals are the main representatives of nutrients needed for maintenance and repair of the body mass. The principal protein components are amino acids, consisting of carbon, hydrogen, nitrogen, and oxygen, sometimes also sulphur and phosphorus. A protein molecule consists of a large number of the individual amino acids linked in various patterns; this is why food chemists can identify so many different kinds of food proteins. Figure 1–1 shows schematically an amino acid sequence and other structural elements of a small pro-

1.5 Nutritional Requirements of a Human Body

Table 1-7 Energy Content of Main Food Components and Some Common Foods[a]

A. FOOD COMPONENTS	ENERGY[b] VALUE/GRAM	
	Kcal	kJ
Protein	4	16.5
Carbohydrate	4	16.5
Fat	9	37.0
Alcohol	7	28.8

B. COMMON FOODS	ENERGY VALUES/100g	
	Kcal	kJ
Milk	66	280
Cheddar Cheese	360	1525
T-bone Steak	165	695
Potato Chips	500	2080
Orange	36	150
Granola	430	1800

[a] Source: Nutritive value of American foods, by C.F. Adams, Agriculture Handbook No 456, USDA, Washington, D.C., and other sources.
[b] One Kcal (Kilocalorie) is the amount of heat energy needed to increase the temperature of 1kg of water by 1°C; One kJ (kilojoule) is the equivalent SI unit in metric system. 1 Kcal = 4'184 kJ.

Table 1-8 Approximate Energy Needs for Various Activities of Healthy Adults[1]

	ENERGY NEEDS		
TYPE OF ACTIVITY	MEN	WOMEN	AVERAGE
	--------kJ/day--------		(kJ/hr)
Sedentary - Light			
Eating, Watching T.V.	9,000–11,500	7,500–10,000	380
Moderate			
Swimming (Pleasure), Bicycling	13,500–15,000	11,500–12,500	1,700
Strenuous			
Forestry Work, Dancing	16,000–17,500	13,000–15,000	2,200
Very Strenuous			
Mountain Climbing, Competitive Sports, Heavy Work	18,500–20,000	16,500–17,500	4,500

[1] Data from various sources.

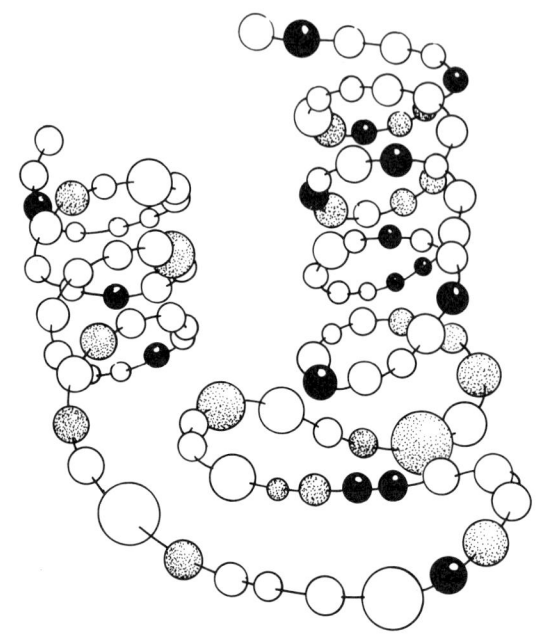

FIGURE 1-1. Schematic illustration of structure of a protein molecule. The dots represent the various amino acids (primary structural elements), linked together in a chain-like sequence (secondary structure), and coiled randomly or in a characteristic helical form (tertiary structure).

tein molecule. Upon digestion, dietary proteins are broken down to their amino acid components that are subsequently used by the body to build "its own" protein molecules as necessary. Eight, or possibly nine, of the twenty-two or so amino acids known to be common in food proteins cannot be synthesized by the body; the remaining ones can if they are not present in the food. The *essential amino acids,* as listed in Table 1-9, must be supplied by the proteins contained in our food. The essential amino acid content of various food proteins is one of the factors used for differentiating their nutritive value. Table 1-10 gives the nutritive values of some common food proteins based on their essential amino acid content. Typical protein intake in the U.S. and Canada is substantially higher than the recommended minimum, and the large variety of good protein sources available to North American consumers virtually assures adequate essential amino acid supply. Possible deficiencies could arise with strict vegetarian diets relying only on a limited number of plant protein sources since many of the plant proteins are lacking in one or more of the essential amino acids.

The indigestible matter contained in foods (principally cellulose, lignin and other components of cell walls found in plant food materials)

1.5 Nutritional Requirements of a Human Body   17

Table 1-9   Amino Acids Essential for Humans

NAME	CHEMICAL STRUCTURE
Isoleucine	CH_3 $\quad\diagdown$ $\qquad\text{CHCHCOOH}$ $\text{C}_2\text{H}_5 \quad\; |$ $\qquad\qquad\text{NH}_2$
Leucine	$\text{CH}_3\diagdown$ $\qquad\text{CHCH}_2\text{CHCOOH}$ $\text{CH}_3\diagup \qquad |$ $\qquad\qquad\text{NH}_2$
Lysine	$\text{CH}_2\text{CH}_2\text{CH}_2\text{CH}_2\text{CHCOOH}$ $\;|\qquad\qquad\qquad\;|$ $\text{NH}_2\qquad\qquad\text{NH}_2$
Methionine	$\text{CH}_3-\text{S}-\text{CH}_2\text{CH}_2\text{CHCOOH}$ $\qquad\qquad\qquad\qquad\;|$ $\qquad\qquad\qquad\qquad\text{NH}_2$
Phenylalanine	$\bigcirc-\text{CH}_2\text{CHCOOH}$ $\qquad\qquad\;|$ $\qquad\qquad\text{NH}_2$
Threonine	$\quad\text{OH}$ $\quad\;|$ $\text{CH}_3-\text{CHCHCOOH}$ $\qquad\qquad|$ $\qquad\qquad\text{NH}_2$
Tryptophan	$\qquad\;\;\text{C}-\text{CH}_2\text{CHCOOH}$ $\bigcirc\!\!\!\diagdown\quad\|\qquad\qquad|$ $\qquad\text{N}\diagup\text{CH}\qquad\;\text{NH}_2$ $\quad\;\;|$ $\quad\;\;\text{H}$
Valine	$\text{CH}_3-\text{CHCHCOOH}$ $\text{CH}_3\diagup\quad\;\;|$ $\qquad\qquad\text{NH}_2$
Histidine[a]	$\text{CH}=\text{C}-\text{CH}_2\text{CHCOOH}$ $|\quad\;\;|\qquad\qquad\;\;|$ $\text{N}\quad\text{NH}\qquad\quad\text{NH}_2$ $\diagdown\;\diagup$ CH

[a] Essential for children and possibly for adults.

is termed dietary fiber or "roughage." Specific requirements for fiber are difficult to establish as the principal role of this "neglected nutrient" is not to supply any biologically active compounds, but rather to provide bulk for proper intestinal activity. Many alleged benefits of high fiber consumption have been suggested, including protection

Table 1-10 Nutritional Quality of Some Common Food Proteins[a]

SOURCE OF PROTEIN	PER[b]	CHEMICAL SCORE[c]
Whole Egg	3.9	100
Fish	3.5	70
Beef Steak	2.3	69
Whole Milk	3.1	60
Soybean	2.3	47
Rice	2.2	56
Peanuts	1.6	43
Wheat Flour	0.6	32

[a] Source: FAO Nutritional studies No 24, Rome, 1970.
[b] Protein efficiency ratio (determined by experiments with rats); g weight gain/g protein received.
[c] Percentage of the most limiting amino acid in a given protein relative to egg protein.

against cancer of the colon, lowering of blood cholesterol, or lowering of blood sugar levels in diabetics. None of these effects has been confirmed by adequate experimental data, and thus manufacturing of foods with high fiber content (supplied principally by wheat bran) may be supported on the basis of consumer demand but not as a dietary need. However, too little fiber, resulting from diets containing mainly foods of animal origin, may lead to medical complications such as constipation or diverticulosis (development and inflammation of weak spots in bowel walls, followed by flatulence, pain or bleeding). According to popular and some scientific literature, the recommended amount of fiber may be 6 to 24 g daily; for comparison, 100g of cooked carrots contains 3.7g, and 100g of green peas 7.8g. Chemically, the main components of the dietary fiber are classified as carbohydrates and thus belong to the group of macronutrients.

Requirements for essential micronutrients are much more complex than those for macronutrients. Today, scientists consider at least fifteen different minerals, some only in trace amounts, and thirteen or fourteen vitamins to be essential. There are differences of opinion as to whether some of these are truly essential (non-synthesizable by humans), but no disagreements exist on their importance for the health and proper function of the body. Some of the more important dietary minerals are listed in Table 1-11 together with recommended intake levels and main bodily effects.

To insure adequate intake of all the needed nutrients, our diet should include a variety of foods, fresh as well as processed. Government health authorities, concerned with the well-being of their citizens, have been issuing various recommendations for proper nutrition. The

Table 1-11 Examples of Important Dietary Minerals and Their Main Functions in the Human Body

ELEMENT	AVERAGE[a] RECOMMENDED DIETARY INTAKE (mg/day)	MAIN EFFECT IN HUMAN BODY
Sodium	2,000	Maintains water balance
Potassium	2,000	Control of nerve impulses
Chlorine	2,500	Forms hydrochloric acid for gastric juices
Calcium	1,200	Building of bones, teeth, proper function of heart
Phosphorus	1,200	Building of bones, teeth, regulation of energy release
Magnesium	300	Conversion of nutrients to energy
Iodine	150	Part of thyroid hormones
Iron	15	Part of oxygen-transfer systems
Zinc	15	Digestion and metabolism of proteins

[a] Compiled by author from various sources.

concept of Recommended Dietary Allowances has been used in the U.S. and Canada to quantitatively express the average daily amounts of individual nutrients required by a healthy body. General nutritional guidelines have been issued recently by professional and legislative bodies such as the Food and Nutrition Board of the U.S. National Research Council, the McGovern's senate select committee on nutrition, or the Department of Health and Welfare, Canada. While the guidelines often differ in specifics, the general thrust of the recommendations is to decrease consumption of fat and simple sugars and to increase intake of starch and other complex carbohydrates. One of the simplest and most useful practical guidelines for everyday nutrition is The Canada Food Guide (issued by the Department of Health and Welfare), which translates the average dietary requirements for individual nutrients into recommended servings of foods containing them. The Canada Food Guide is reproduced in Figure 1-2. An American version, the Daily Food Guide, has been in use since 1957.

1.6 NUTRITIVE VALUE OF PROCESSED FOODS

The proximate composition of raw and processed foods is one useful indication of their nutritive value. All five groups of the proximate components—including water—are important for human nutrition.

FIGURE 1-2. The Canada Food Guide. (Source: Health and Welfare Canada.)

(Canada's Food Guide)

Eat a variety of foods from each group every day

Energy needs vary with age, sex and activity. Foods selected according to the guide can supply 1000-1400 calories. For additional energy, increase the number and size of servings from the various food groups or add other foods.

milk and milk products

Children up to 11 years	2-3 servings
Adolescents	3-4 servings
Pregnant and nursing women	3-4 servings
Adults	2 servings

Skim, 2%, whole, buttermilk, reconstituted dry or evaporated milk may be used as a beverage or as the main ingredient in other foods. Cheese may also be chosen.

Examples of one serving
250 ml (1 cup) milk, yoghurt or cottage cheese
45 g (1½ ounces) cheddar or process cheese

In addition, a supplement of vitamin D is recommended when milk is consumed which does not contain added vitamin D.

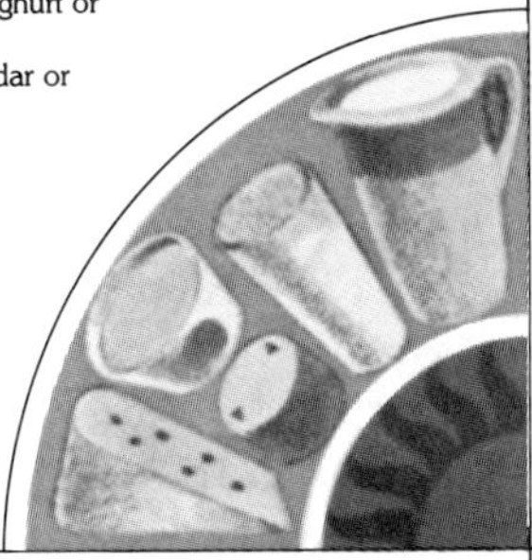

meat and alternates
2 servings

Examples of one serving
60 to 90 g (2-3 ounces) cooked lean meat, poultry, liver or fish
60 ml (4 tablespoons) peanut butter
250 ml (1 cup) cooked dried peas, beans or lentils
80 to 250 ml (⅓-1 cup) nuts or seeds
60 g (2 ounces) cheddar, process or cottage cheese
2 eggs

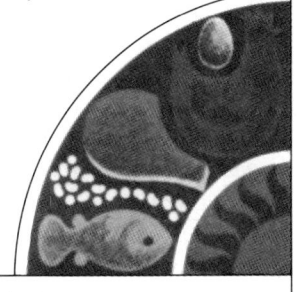

bread and cereals
3-5 servings

whole grain or enriched. Whole grain products are recommended.

Examples of one serving
1 slice bread
125 to 250 ml (½-1 cup) cooked or ready-to-eat cereal
1 roll or muffin
125 to 200 ml (½-¾ cup) cooked rice, macaroni, spaghetti

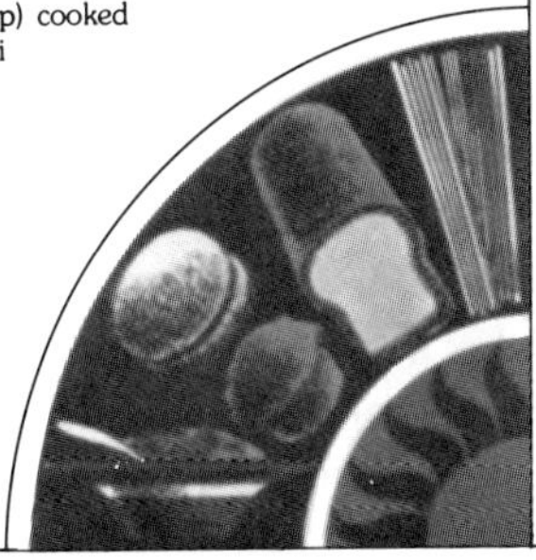

fruits and vegetables
4-5 servings

Include at least two vegetables.

Choose a variety of both vegetables and fruits — cooked, raw or their juices. Include yellow or green or green leafy vegetables.

Examples of one serving
125 ml (½ cup) vegetables or fruits
125 ml (½ cup) juice
1 medium potato, carrot, tomato, peach, apple, orange or banana

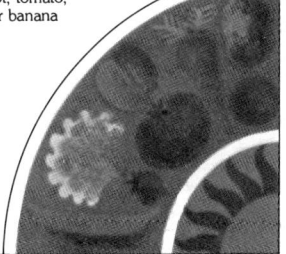

FIGURE 1–2. (*Cont.*)

However, much too often consumers only look at the protein and vitamin content of foods as indicators of their "healthfulness," or nutritive value, and at the sugar and fat content as negative components.

Since micronutrients are present in foods in very small amounts, their content must be determined separately from the proximate analysis. The values are usually expressed in milligrams per unit of serving, or, as shown in Table 1-12, in mg/100g food. Comprehensive tables listing several hundred fresh as well as processed foods are available from Health and Welfare Canada or the United States Department of Agriculture (USDA).

In Table 1-12, data for recommended daily intake of the selected micronutrients are also enclosed to illustrate that it is very easy to obtain the micronutrient requirements by eating a varied diet including home-cooked or industrially processed foods. It can also be seen that some of the often-condemned industrial foods, such as canned vegetables, are a good source of some of these nutrients; that none of the food items listed—fresh or processed—contains all the nutrients required; and that losses of nutrients in food processing are not excessive. In many cases, home cooking results in the same or even greater losses, depending on the skill of the cook and the process used. For example, when the vitamin C content of hot, dinner-ready green peas is compared to uncooked green peas from a garden, it matters very little whether the steaming bowl was prepared from fresh, frozen, or canned peas. The percent of retention as compared to the fresh product will not differ greatly, with the fresh peas retaining slightly more (about 45 percent) than the frozen (40 percent) or the canned (30 percent) products.

Most food processing operations, including home meal preparation and prolonged storage of unprocessed agricultural materials on the farm, in a warehouse, in a health food store or at home, will result in losses of certain micronutrients. Fresh produce eaten directly from the garden is unquestionably a better source of vitamin C than its processed or cooked counterparts. The various food processing operations used by the industry are usually a compromise between technological feasibility, economic feasibility or effectiveness, and minimum destruction of nutrients. It would be pointless to grind all our flour on stone mills if the mills could not be industrially operated and/or the flour would be too expensive for most people to buy.

Processing operations may affect food nutrients in two principal ways. Macronutrients as well as micronutrients may be physically removed from the food. Examples include removal of fiber and some vitamins in grinding wheat kernels for flour, leaching of water-soluble minerals and vitamins in processing of vegetables, or removal of fat-soluble vitamins and fat in separation of cream from milk. The second route involves deactivating the biological functions of unstable vitamins by means of a deactivating agent such as heat, light, or oxygen

Table 1–12 Approximate Micronutrient Content[a] of Selected Food Items

	PEAS		ORANGE JUICE		SALMON		MILK		
Micronutrient	Raw	Canned	Fresh	From Frozen Concentrate	Fresh, Broiled	Canned	Fresh (Pasteurized, 3.5% fat)	Fermented (Buttermilk, 0.1% fat)	Dietary Requirements[b] For Adults (mg/day)
				mg/100 g					
Thiamine	0.3	0.1	0.1	0.1	0.2	—	—	—	1.2
Riboflavin	0.1	—[c]	—	—	0.2	0.2	0.2	0.2	1.5
Niacin	2.9	0.8	0.4	0.3	9.9	7.4	0.1	0.1	18.0
Vitamin C	27.0	8.0	49.6	47.3	—	—	0.9	1.0	60.0
Calcium	26.2	26.2	10.8	10.0	—	—	118.0	122.0	1200
Iron	1.9	1.9	0.2	0.1	1.2	1.0	—	—	15
Phosphorus	116.8	77.0	16.8	16.8	417.3	288.2	90.8	95.7	1200
Potassium	318.4	96.6	198.4	198.4	446.4	339.1	143.4	141.1	2000

[a] Adapted by author from "Nutritive Value of American Foods", by C.F. Adams, Agric. Handbook, No. 456, USDA, Washington, D.C.
[b] From Tables 1–5 and 1–11.
[c] Less than 0.1 mg/100 g.

used in the process or present in storage. Destruction of some thiamine in braised beef steak or deactivation of riboflavin in milk stored under direct light are two illustrations of the sensitivity of certain food vitamins. In many cases, the effects of food processing on vitamin retention may be insignificant in the total context of the particular process or the given nutrient availability from other sources. As discussed in Chapter 6, pasteurization of milk to insure complete removal of pathogenic microorganisms destroys a significant percentage of vitamin C in milk. However, we do not drink milk to receive our vitamin C needs, since in absolute terms, milk is a very poor source of this particular vitamin.

Many processes may affect only some nutrients while improving the digestibility or bioavailability of others. Perhaps a slightly absurd but fitting example is the process of hand deboning of meat. Hand deboning removes one of the best sources of dietary calcium—the inedible bone. Nutritionally, this is of no consequence as we get dietary calcium from milk and other dairy products. Other nutrients in the beef steak such as thiamin, iron, etc., are not removed in the deboning process; in the subsequent cooking step, some thiamin is lost but the protein is made more digestible. Improved digestibility by gelatinization of starch in the bread-baking process is another example of positive effect of heat on nutrients.

In general, North Americans have available a large variety of fresh as well as processed foods. Thus, the concerns with nutrient losses in processed foods are often academic. Anyone eating a varied, adequate diet containing foods from the main four groups of fruits and vegetables, bread and cereals, meat and fish, and milk and milk products, should be more than adequately nourished.

1.7 INDUSTRIAL FOOD PROCESSING AND HEALTH FOODS

As will be shown throughout this book, food processing today is a complex industrial activity. Because the basic principles of these processes are not widely understood, there appears to be a widespread belief in the general population that industrial food processing destroys nutritive values. Worse yet, accusations regarding "chemicals" or even "poisons" in our foods are becoming more and more common. In addition, scientific disagreements and controversies in the sphere of human nutrition, such as the polyunsaturated vs. saturated fat and cholesterol debate (see Chapter 5), are often reported by the press in an incomplete or distorted fashion, adding to the consumers' confusion. One indication of the levels of misinformation among the general public is the continued growth of the "organic food—health food" industry. These products, although no more nutritional than their industrially processed counterparts, are sold in special stores, often at much higher prices than in the supermarkets. However, because it is claimed that no "artificial chemicals" were used in the growing or processing of "organic foods," the consumer is led to believe that the natural component chemicals of the produce are somehow better.

The "health foods" movement is a characteristic sign of a modern affluent society that has forgotten about famine, and can afford to pay a premium for simple nutrients. It is also a result of significant advances that medicine has made in recent years. Since the threat of killer diseases such as bubonic plague, tuberculosis, and smallpox have

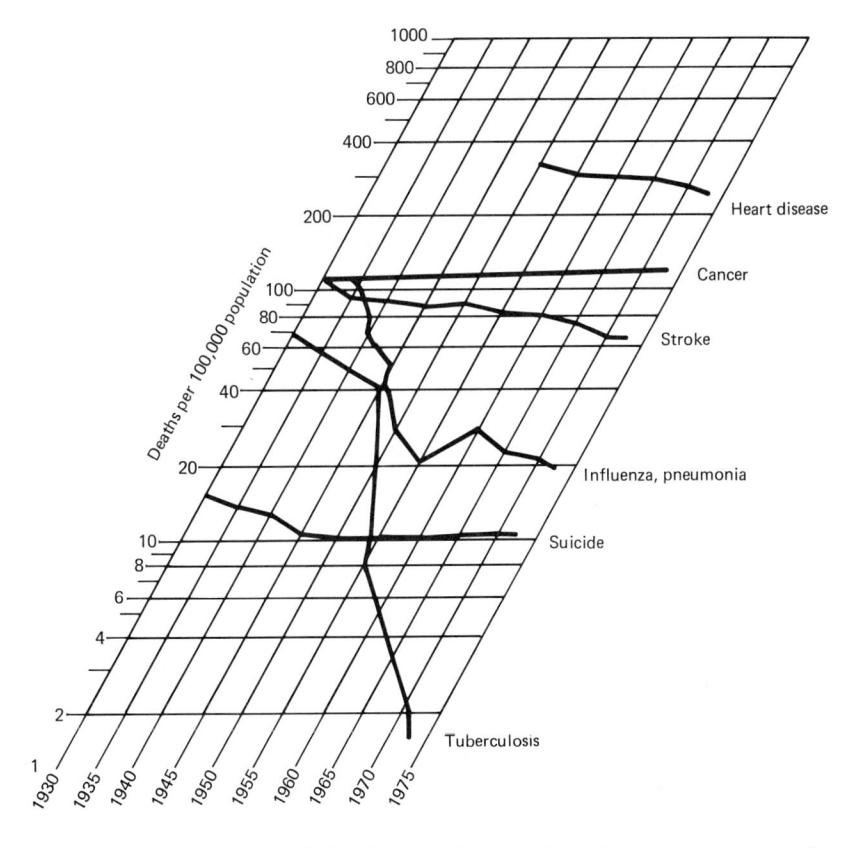

FIGURE 1-3. Age-adjusted death rates from selected causes. Reprinted by kind permission from Food Technology 1978, 32 (9):51. Copyright © by Institute of Food Technologists.

been virtually eradicated, the main health concern of general populations has shifted towards degenerative diseases, especially cancer and cardiovascular disease. The causes of these diseases are still poorly understood. But because of virtual elimination of the previous killer diseases, degenerative diseases have become, relatively speaking, the prime killers although their absolute incidence per capita has been steadily decreasing, as shown in Figure 1-3.

The advances of modern nutritional science have made it possible to recommend optimum diets for health maintenance. However, this does not mean that diet can be relied on to eliminate all disease. There is no such thing as preventative nutrition for immortality. According to the Food and Nutrition Board of the United States National Research Council[1]... "Sound nutrition is not a panacea. Good food that provides appropriate proportions of nutrients should not be regarded as a poison, a medicine, or a talisman. It should be eaten and enjoyed."

1.8 CONTROL QUESTIONS

1. Explain the terms "proximate composition" and "gravimetric analysis."
2. If potatoes, on the "as is" basis, are 80% moisture, 2% protein, and 18% carbohydrate, what is the proximate composition of the potatoes on a dry basis?
3. Why is the "total carbohydrate content" of a food often determined by difference?
4. What is the nutritive value (in terms of food energy only) of a medium-done 10-oz. sirloin steak (10 oz. is approximately 300g) which has been analyzed in a laboratory to contain 45% H_2O and 23% protein? Do not forget to estimate the fat content of the steak for the determination.
5. What kinds of macronutrients and micronutrients might be affected by a food processing operation involving keeping a product (e.g. diced carrot) in 70°C hot water for 15 min.? What would be the likely effect(s) on the individual nutrients?
6. Among the various products sold in health food stores are vitamin supplement pills. Do you think that these products are important for North American consumers? Explain.

---

[1] *Toward Healthful Diets*, (Washington, D.C.: National Academy of Sciences, 1980), p 17.

7. One of the most dreaded diseases for seamen exploring the Canadian North was scurvy. Why was scurvy so prevalent in these men?

Suggested Reference Books—
Foods, Food Composition and Nutrition

Adams, C. F. 1975. *Nutritive Value of American Foods* (Agric. Handbook No. 456). Washington, D.C.: USDA.
Bender, A. E. and Nash, T. 1979. *Pocket Encyclopedia of Calories and Nutrition*. New York: Simon and Schuster.
Bogert, L. J., Briggs, G. M., and Calloway, D. H. 1973. *Nutrition and Physical Fitness*. Philadelphia, Pa.: W. B. Saunders.
Gates, J. C. 1976. *Basic Foods*. New York: Holt, Rinehart and Winston.
Harris, R. S., and Karmas, E. 1975. *Nutritional Evaluation of Food Processing*. Westport, Conn.: AVI Publishing Co.
Health Protection Branch. 1976. *Dietary Standard for Canada*. Health and Welfare Canada, Ottawa: Bureau of Nutritional Sciences.
Health Protection Branch. 1977. *Nutrient Value of Some Common Foods*. Health and Welfare Canada, Ottawa: Supply and Services.
Meyer, L. H. 1975. *Food Chemistry*. Westport, Conn.: AVI Publishing Co.
Wenck, D. A. 1981. *Supermarket Nutrition*. Reston, Va.: Reston Publishing Co.
White, P. L. and Selvey, N. editors. 1974. *Let's Talk About Food*. Acton, Mass.: Publ. Sciences.

2

Microorganisms in Foods

2.1 FORMS OF MICROBIAL LIFE

Raw materials for human foods come predominantly from two biological kingdoms—the *plants*, and the *animals*. Living organisms belonging to the third biological kingdom, the *protista* (characteristically organisms of microscopic size and low level of morphological complexity) do not supply appreciable amounts of raw food material. However, these simple, often unicellular *microorganisms* have a pivotal influence on all aspects of food production, processing, distribution and consumption. An appreciation of interactions between microorganisms and food materials is essential to understand the diverse food manufacturing processes used by the modern food processing industry.

The study of microorganisms is complicated because of their very small size, often below the limits of visibility to the naked eye (Figure 2-1). The existence of microorganisms was assumed by Greek and Roman scholars, but it was not until 1676 when Dutch scientist Antonie van Leeuwenhoek discovered, by microscopic observations, the widespread existence of bacteria. It took another 200 years before Pasteur in his historical experiments proved that all putrefactive spoilage of biological matter (including food) is caused by microorganisms. Today, prevention of undesirable microbial activity is the primary task of most food processing and preservation techniques.

The main forms of microorganisms important to the study of foods are bacteria, yeasts and molds. Other types of microscopic organisms (ricketsia, viruses, protozoa, parasitic larvae, etc.) may also play an

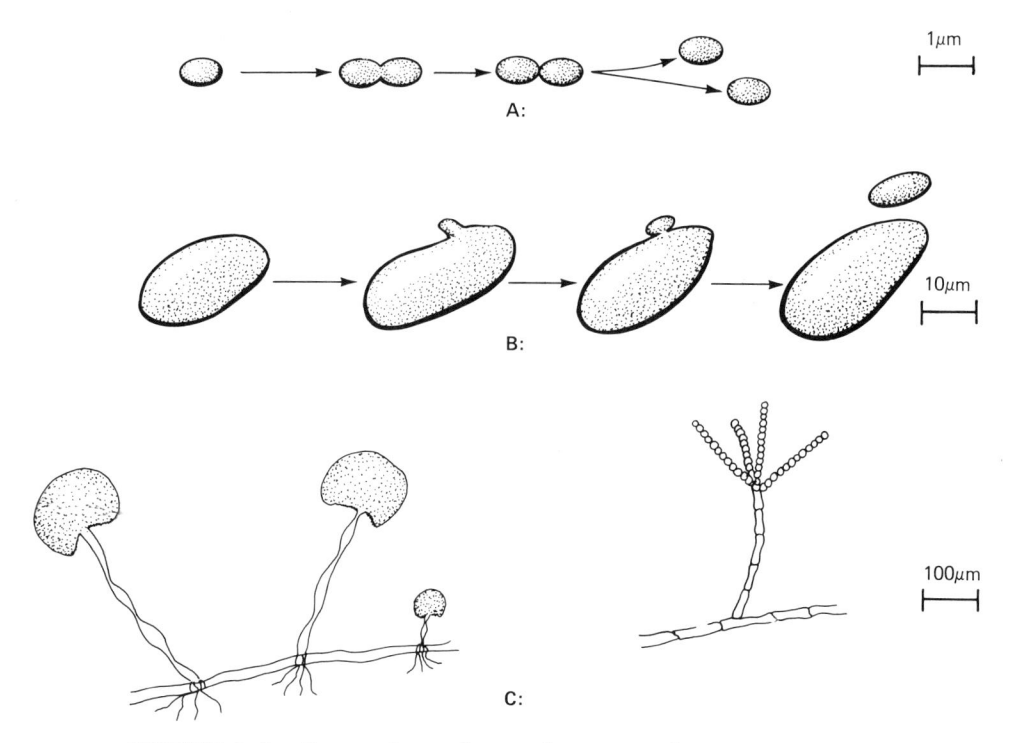

FIGURE 2-1. Comparison of reproductive mechanisms and morphological differences in bacteria (A), yeasts (B), and molds (C).

important role in certain specific cases of food safety or food fermentation processes.

Bacteria, one of the simplest forms of microbial life, are unicellular organisms of approximately 1 µm in size. They reproduce by binary fission, whereby a bacterial cell absorbs nutrients from its environment and transforms them into microbial biomass. Eventually, this leads to the single cell splitting into two new ones. When such a propagation of life occurs simultaneously in many individual vegetative cells present in a given food, we speak of the phenomenon of *microbial growth*. Food poisoning, food spoilage, or production of certain foods by fermentation are all results of microbial growth.

During the vegetative period the microorganisms utilize much the same nutrients from various biological materials as humans do. In the process of converting the available nutrients into new microbial cells, many different compounds are formed by the numerous biochemical processes of the microbial metabolism. Some of the new compounds become a part of the new biomass. Many other products of the breakdown of the organic matter, as well as products of the microbial synthesis mechanisms, are "excreted" by the cells as metabolic byproducts. The bitterness in spoiled milk, the toxin in moldy peanuts, or the delicate flavor of fermented cabbage (the sauerkraut) are all caused by substances produced by microorganisms during their vegetative growth.

Bacteria can exist either as a *vegetative cell* or as a *spore*. Vegetative cells represent the active form of bacterial life while spores, in their dormant state, do not show any signs of metabolic activity. Under certain favorable conditions a change of a spore into a vegetative cell may occur in a process called germination.

Only a few types of bacteria have the ability to form spores, which may be viewed as a survival mechanism for injurious environmental conditions, rather than a mechanism of reproduction. Spores are usually highly resistant to heat, acid, or other conditions commonly found to prevent growth or even kill vegetative cells. Food sterilization processes (Chapter 10) are particularly concerned with complete inactivation of the most heat-resistant spores, whose survival and possible delayed germination could lead to spoilage problems or toxicity hazards. Some of the important spore-forming and non-spore-forming bacteria related to food processing are listed in Table 2–1.

Yeasts and *molds* also produce spores. However, these are not highly resistant to environmental effects since they are a mechanism of reproduction rather than survival. Both yeasts and molds, which are much larger organisms than bacteria, also reproduce by a cell division mechanism similar to that of plant cells. Figure 2–1 shows schematically some of the morphological and reproductive differences between bacteria, yeasts and molds.

Vegetative yeasts and molds generally show lower tolerance to heat than some types of bacteria, while extremes in acidity and osmotic pressure, caused by the presence of acids, salts and sugars in the aqueous environment, are less inhibitory to yeasts and molds than to bacteria. Certain bacteria, yeasts and molds are very useful in the manufacture of common foods (examples described elsewhere in this text include cheese, wine, fermented sausage, bread and other traditional delicacies), while many other microorganisms are undesirable as food contaminants since their growth will result in food spoilage.

2.2 MICROBIAL GROWTH AND FOOD SPOILAGE

Microbial spoilage of food is caused by active *growth* of vegetative forms of microorganisms. The rate of microbial multiplication in a favorable environment is unusually rapid; the doubling time for some bacteria may be less than 20 minutes under optimal conditions. This is one of the characteristic aspects of microbial spoilage. Other causes of

Table 2-1 Some Important Bacteria Related to Food Processing

GROUP	EXAMPLE	IMPORTANCE TO FOOD PROCESSING
a) Spore-Forming Bacteria		
Clostridium	C. thermosacharolyticum	Spoilage of canned foods
	C. botulinum	Toxicity of under-processed canned foods
Bacillus	B. stearothermophilus	Used as indicator organism for antibiotics in milk
b) Non-Spore-Forming Bacteria		
Pseudomonas	P. putrefaciens	Psychrophilic spoilage and bitterness in milk
Staphylococcus	S. aureus	Food intoxication
Salmonella	S. typhimurium	Food-borne infection
Coliform group	E. coli	Indicator of unsanitary processing
Lactobacillus	L. bulgaricus	Used for production of fermented dairy products

food spoilage, such as chemical deterioration or enzymatic reactions, proceed usually at a slower pace because the causative agents are at a fixed level and do not multiply like the microorganisms.

The growth of microorganisms is influenced by environmental conditions. There are many different kinds of microorganisms with specific requirements for temperature, pH, oxygen, water and nutrient availability, and other factors. Some of these requirements may not be satisfied by human food materials; this is why only certain species of the many microorganisms found in our environment are important for foods. Table 2-2 indicates the main factors which influence the growth of various organisms. The responses of individual types of microorganisms to these factors may vary substantially. Thus, with respect to temperature, most microorganisms can be classified as *mesophilic*, as their optimum growth will occur at the 30 to 34°C temperature range. Other

Table 2-2 Effects of Environmental Factors on Growth of Microorganisms

FACTOR	TYPE OF MICROORGANISM	GROWTH REQUIREMENTS
Temperature	Psychrophilic	Will grow at refrigerated temperatures
	Mesophilic	20–36°C
	Thermophilic	40–55°C
Oxygen	Aerobic	O_2 required
	Anaerobic	Absence of O_2 required
Water[a]	Bacteria	$a_w = 0.9$ or higher
	Yeasts	$a_w = 0.88$ or higher (some will grow at $a_w = 0.7$)
	Molds	$a_w = 0.8$ or higher (some will grow at $a_w = 0.6$)
Acidity	All	Approximate range of tolerance pH 4–pH 8; specific for specific organisms; some yeasts and molds very acid-tolerant
Nutrients	All	Wide range of requirements

[a] The effect of water is usually expressed in terms of water activity (a_w) as defined in Section 2.3; see also Table 2-4.

species, denoted as *thermophilic*, require much higher temperatures of up to 55°C or even more to grow well. On the other hand, there are some cold tolerant organisms called *psychrophiles* or *psychrotrophs*, that are capable of growth at 10°C or lower.

Oxygen, one of the principal growth factors for many living organisms, may be inhibitory for some bacteria. The majority of microorganisms are *aerobic* since they require oxygen for their growth, while obligate *anaerobes* will grow only in complete absence of oxygen. Some anaerobic bacteria require small amounts of oxygen to be present and these are sometimes termed *microaerophilic*.

Responses of various microorganisms to varying concentrations of acids, salts, sugars and other compounds that may interact with the availability of water (see Section 2.3) are highly specific for each species. Many organisms, especially certain yeasts and molds, can tolerate relatively high amounts of acids that are present in some fruits and fruit juices with a pH of 4.0 or less. *Osmophilic* organisms will grow in the conditions of relatively high osmotic pressure caused by increased concentrations of salts and sugars. In general, availability of free water is one of the key factors for all microbial growth; as an example, properly dried foods will not spoil from microbial causes.

There are many highly specific nutrients and other growth factors required by the individual microorganisms. Absence of a single essential nutrient, or a presence of an inhibitor, can completely eliminate growth of a given organism in otherwise suitable conditions where many other organisms may thrive.

During their growth, microorganisms usually produce various organic acids as the end-product of their metabolism. Many microorganisms also produce CO_2 gas and sometimes other gases in addition to the acid production; these are called *heterofermentative* organisms as opposed to the *homofermentative* ones which produce acid only. In many cases of microbial growth, the metabolites (such as the lactic acid produced during yogurt fermentation) may prevent further growth of the producing organism as well as deter many other spoilage organisms that may be present.

Food preservation techniques are based on manipulation of one or more of the growth factors to prevent the microbial contaminants from active multiplication. Because of the wide diversity in the growth requirements of many organisms, it is very seldom that all microbial growth can be stopped by a given process. Typical examples in this regard are refrigeration, which eliminates growth of most microorganisms (but not psychrophiles), or pickling, which will retard growth of all but the acid-tolerant, osmophilic organisms. There are very few food processing techniques which actually eliminate all viable forms of microbial life from the food, which then must be properly packaged to avoid recontamination from the environment. Canned or UHT-sterilized foods (see Section 10.4) are the only major group of food products whose long-term shelf life is based on killing of all microorganisms present rather than on preventing them from active growth.

2.3 WATER ACTIVITY

One of the most important growth factors required by all microorganisms is water. Two of the three main long-term food preservation processes, drying and freezing (Chapter 10), eliminate water as an available nutrient, thus preventing microbial growth. Other food processing techniques are based on altering the composition of food materials or final food products to decrease the availability of the constituent water to microorganisms. The salting of fish and the method used for the manufacture of jams are typical examples of this approach based on controlling the activity of the water molecules present in food.

Two foods of the same moisture content may exhibit different susceptibility to microbial spoilage. As shown in Table 2–3, a sausage and a

Table 2-3 Composition[a] of a Sausage and a Salted Fish and Their Estimated Water Activity

COMPOSITION	SALTED CODFISH	POLISH SAUSAGE
	──── % Weight ────	
Water	52.4	53.5
Protein	29.0	15.6
Fat	1.0	25.8
Carbohydrate	0.0	1.1
NaCl	14.0	1.0
Estimated a_w	0.90	0.99

[a] Adapted from *Nutritive Value of American Foods*, C.F. Adams, Agriculture Handbook No. 456, USDA, Washington, D.C.

salted codfish contain about the same amount of water. However, the sausage will probably spoil faster than the cod, because the water molecules in the fish are "tied up" by the salt and thus are much less available to the microorganisms. As shown by the example, compositional data about a food, especially its moisture content alone, do not provide sufficient information regarding the availability of water for microbial growth. Informed judgements about susceptibility of foods to microbial spoilage must be based on knowledge of the physicochemical *activity* of the water present.

Water activity, symbolically a_w, is related to the state of the water molecules in the food. Water which is "free" can evaporate from food easily. Thus a_w is measured as a dimensionless *ratio* of pressure of the water-vapor present in air immediately surrounding the food sample in equilibrium conditions, to that of vapor pressure above pure water:

$$a_w = \frac{P_{H_2O} \text{ (food)}}{P_{H_2O} \text{ (water)}} = \frac{P_f}{P_o} \qquad \text{(Equation 2-1)}$$

As the values for P_f can range from 0 (for a completely dry food) to P_o, the pressure above pure water, the range of a_w values will be from 0.0 to 1.0.

In experimental measurements of a_w of foods only the P_f has to be found as P_o, the water vapor pressure in air saturated with H_2O, is dependent only on temperature as shown in Figure 2-2. The amount of water vapor (and its corresponding pressure P_f in air surrounding a food in an enclosed environment) will be determined by the amount of solutes (M_s) like sugars, salts and other components of this food that are dissolved in the water present:

$$P_f = P_o \cdot \frac{M_{H_2O}}{M_s + M_{H_2O}} \qquad \text{(Equation 2-2)}$$

where M_s and M_{H_2O} represent the amounts of a solute s and of H_2O in *moles*. The Equation 2-2 is an expression of *Raoult's law* that is exactly obeyed by ideal solutions only. In many food applications, however, the approximations obtained by using Raoult's law are sufficiently accurate. As can be seen, soluble substances of small molecular weight such as salts will have a greater effect on the lowering of vapor pressure P_f than large molecular weight solutes when taken at equal mass concentrations. Solutes of very large molecular weights such as proteins, or components that are not soluble in water like fat or bones, will have practically no effect on P_f. Thus, the codfish containing 14 percent salt will have much lower water activity than the sausage which only contains 1 percent salt and no other small solute molecules that could effectively reduce the a_w. Data for water activity of various common foods are presented in Table 2-4.

The response of individual bacteria, yeasts or molds to changes in water activity is greatly variable. In general, no microbial growth can be expected at a_w below 0.6; no major spoilage bacteria will tolerate a_w below 0.9. An approximate a_w tolerance scale for microorganisms is shown in Figure 2-3.

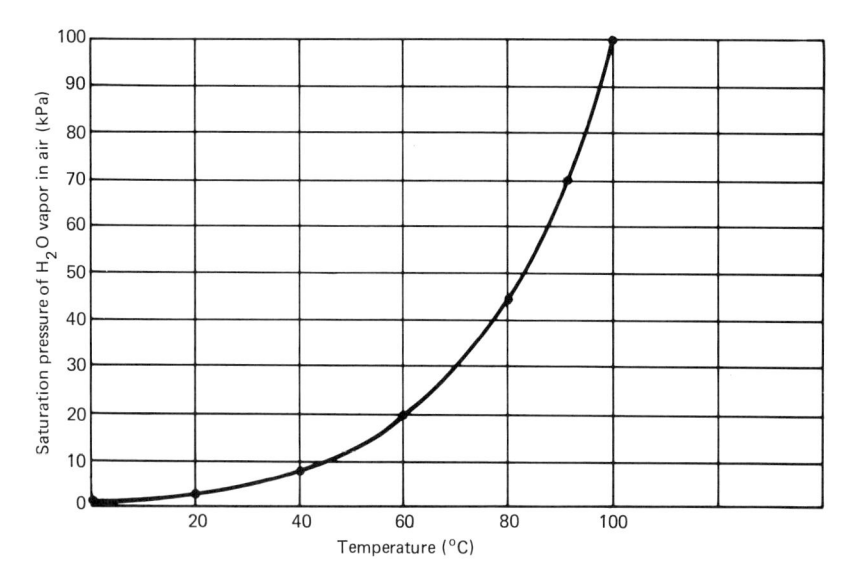

FIGURE 2-2. Effect of temperature on saturation pressure of water vapor in air (equivalent to boiling point of water at a given absolute pressure). 100 kPa = 1 bar (approx. 1 atm) = 14.5 psi.

Table 2–4 Water Activity Data for Some Common Types of Food

PRODUCT	AVERAGE MOISTURE CONTENT (%)	WATER ACTIVITY (a_w)
Fresh meats, fruits, vegetables	90–60	1.00–0.99
Aged cheddar cheese, salami, blue cheese, orange juice concentrate	50–40	0.98–0.95
Baked goods, processed cheese, bacon	40–20	0.95–0.90
Jams, dates, sweetened condensed milk, Parmesan cheese	30–20	0.90–0.75
Honey, pasta products	20–10	0.60–0.50
Dried milk powder, dried potato, dry mixes	5–2	0.30–0.20

Industrial processes for manufacturing food products called intermediate moisture foods (IMF) have been developed relatively recently as a side–benefit of the U.S. space exploration program. The technology of these products, based on the a_w control by compounds called *humectants*, is discussed Chapter 9.

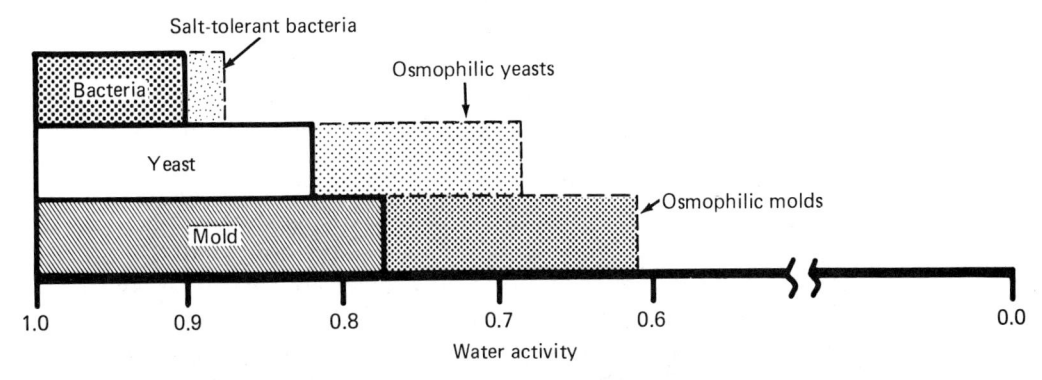

FIGURE 2–3. Water activity tolerance scale for major groups of microorganisms.

2.4 FOOD POISONING OR FOOD-BORNE ILLNESS

Among the great many types of microorganisms known to be in existence, there are only relatively few that may interfere with human health. Some of these disease-causing *pathogenic* microorganisms may be found in—and transmitted by—raw or processed foods. From the standpoint of food safety, the *presence* of any pathogen is undesirable, although in some cases, these microorganisms must grow before the contaminated food becomes a potential cause of food-borne illness, or, in lay language, "food poisoning." There are two distinctly different types of food-borne illness: *food intoxication* and *food infection*.

The growth of pathogenic microorganisms in food is generally required for a case of food intoxication to occur. This type of food poisoning is caused by a food in which a specific pathogenic organism had grown and produced, as a result of its metabolism, a chemical compound toxic to the human body. The most dangerous food poisoning organism is *Clostridium botulinum*, an anaerobic spore former that may grow in improperly sterilized (canned) food if the spores survive the canning process. *C. botulinum* produces a strong toxin which, if ingested in only a very small dose of 1 ng or less, can be fatal. Another food poisoning organism which is less lethal is *Staphylococcus aureus*. Its toxin is much less harmful but, contrary to the botulotoxin, the staphylococcal toxin is heat–stable so the foods infested by *S. aureus* cannot be made safe by heating. In general, food intoxication cases are characterized by a short onset time, usually two to eight hours from ingestion of the food.

In contrast to food intoxication, food-borne infection is caused by ingestion of sufficiently high numbers of a given disease-causing organism present in a food. The onset of the illness may occur twenty-four hours or more after eating the contaminated food, and the length and severity of the incapacitation may vary substantially, depending on health status and other personal specificities of the patient. In the case of food infections, the pathogen grows in the human body rather than in the food. Typical examples of food infection organisms are some members of the *Enterobacteriaceae* family, especially those belonging to the genus *Salmonella* (the causative agents of typhoid fever and similar intestinal diseases) or *Shigella* (causing various bacillary dysenteries). Although many foods have been implicated in outbreaks of Salmonellosis, the most common carrier appears to be poultry and other raw meats especially if left unrefrigerated for extended periods of time. Another example of a food-borne infection is the Q-fever caused by a rickettsia organism (a parasitic organism falling in between bacteria and viruses) *Coxiella burnettii*, found sometimes in raw cow's milk (see Chapter 6). Several other microorganisms were identified relatively recently as possible causative agents of food-borne illness such as *Yersinia enterocollitica, Campylobacter jejuni,* or *Bacillus cereus*. A list of the most important food pathogens with their potential health hazards is provided in Table 2-5.

Assurance of microbial safety is one of the most important aspects of industrial food quality control. While some of the food pathogens need only to be present in very small numbers to make a food unsafe, checking all foods for the presence of all pathogens would be impractical. Microbiological quality control procedures are often based on a concept of *indicator organisms* (compare with using the enzyme peroxidase in blanching as described in Chapter 4). A typical indicator organism is *Escherichia coli* and other members of the *coliform* group. Presence of coliforms, which are organisms associated with intestinal tract microflora, is usually an indication of unsanitary handling or processing procedures. Coliforms as a group are easy to detect, are non-pathogenic (and thus not dangerous to the health of laboratory personnel), are associated with the source of potential pathogens, and, by experience, are present in much greater numbers than the pathogens. Thus, the presence of coliforms in a food sample may indicate poor sanitation or even the possible presence of pathogenic microorganisms.

Table 2-5 Selected Pathogenic Organisms That May Be Associated With Foods

	POTENTIAL HEALTH HAZARD	FREQUENCY
a) Food Intoxication		
Clostridium botulinum	Severe poisoning, often death; toxin heat labile	Very rare
Clostridium perfringens	Diarrhea, cramps, death rare (little toxin produced in the food)	Very common
Staphylococcus aureus	Stomach cramps, diarrhea, vomiting; toxin heat stable; death rare	Common
b) Food-borne Infections		
Salmonella spp.	Fever, diarrhea, vomiting may last several days, occasional death in young or aged	Very common
Campylobacter jejuni	Abdominal pain, bloody diarrhea, generally fever; vomiting and dehydration rare	Rare
Yersinia enterocolitica	Diarrhea or inflammation of lymph nodes of the intestine; can be fatal	Rare

Common procedures used for detection of coliforms and other bacteria are introduced in section 2.5.

Food intoxication and food-borne infections are relatively rare in North America according to medical statistics. Although it is likely that many cases of mild food-borne illness ("24-hour stomach flu") are unreported, the safety record of the modern food processing industry is outstanding. Most of the recently reported cases of severe food poisoning have originated from home-processed foods. Typical situations that may result in food–related epidemics involve unheated perishable foods (salads, cream pastries, cold chicken) that may have been handled in unsanitary conditions. Cutting ready-to-eat chicken with equipment used for preparation of the raw material can be one of the most dangerous kitchen situations. *Staphylococcus* poisoning, often associated with hand-made salads, cole slaw, cold sliced ham, etc., may be related to the health status and conscience of the worker since *S. aureus* is characteristically present on human skin and especially in exposed cuts, healing wounds and other skin disorders. The nature of the restaurant industry makes it especially vulnerable to the danger of food-related epidemics if proper food preparation and serving procedures are not used. From this standpoint, some of the fast-food chains are probably the "safest" places to eat since the most modern food preparation is usually the norm.

2.5 DETECTION AND ENUMERATION OF MICROORGANISMS

The presence of microbial life in all foods—with the possible exception of sterilized products—is inevitable. Because of the size and the ubiquitous nature of microorganisms, special laboratory techniques must be used in determining the microbial populations in foods.

Microbial content is measured in *numbers* of viable organisms per weight unit (usually a gram) of a food. When total numbers of all microorganisms present in a food are of interest, we speak of *total count*. Specific types of microorganisms (such as coliforms, yeasts or molds) are determined by various techniques that will detect only the specific kind for which the test has been designed. Since microorganisms are invisible but multiply rapidly, microbiologists must use indirect techniques based on detection of the desired microorganisms in a carefully formulated nutrient medium, into which a precisely measured amount of food is thoroughly dispersed. The nutrient medium, nutrient *"agar,"* is propagated for 24–48 hours under carefully controlled temperature conditions. The agar medium solidifies after the food has been dispersed, and each viable microbial cell will produce a separate colony of

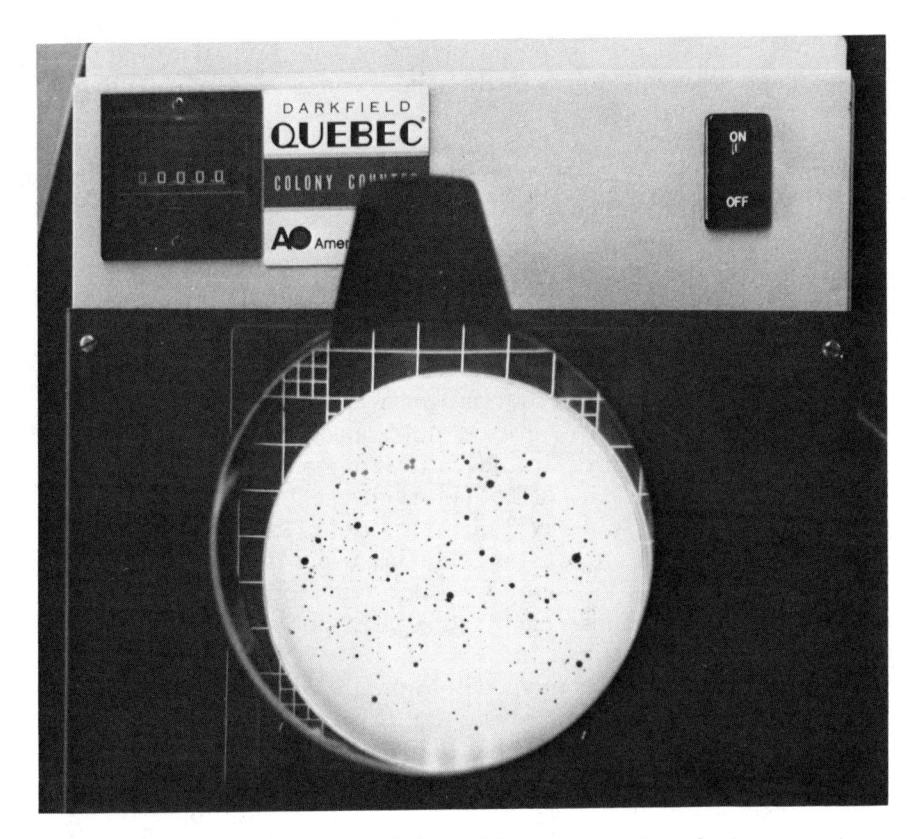

FIGURE 2–4. Incubated petri dish used for enumeration of microorganisms in food.

cells, visible with a magnifying glass or by the naked eye (Figure 2–4). Counting the colonies will provide the required information about the microbial cells or "colony forming units" (CFU) originally present in the food. Other methods, including direct observation under a high magnification microscope, or measurement of microbial growth by development of turbidity in a liquid medium, may be also useful in some conditions. The majority of routine laboratory quality control techniques are based on colony counting.

Typical levels of microbial populations of food vary from essentially germ-free, sterilized foods, to counts as high as 10^7 - 10^8 CFU/gram of food. Many raw materials containing hundreds of thousands of microorganisms per gram are considered of good microbial quality. The North American raw milk standard of 65,000 CFU/g is the envy of all European dairy processors who receive raw material frequently containing more than 1 million CFU/g. As a general approximation, total counts of 10^6 - 10^7 CFU/g of food may be indicative of impending spoilage problems unless the food has been produced by deliberate action of microorganisms.

2.6 FOOD FERMENTATIONS

Together with the interest areas of food safety and food spoilage, the use of microorganisms in food fermentations and other manufacturing processes constitutes the third major segment of food microbiology. Many traditional food processing techniques discussed later in this text rely on the action of microorganisms producing the desired flavors (beer, wine, sauerkraut, cultured dairy products), texture (bread, yogurt) or other specific desirable features of many food products (eyes of Swiss cheese, blue veins of Roquefort cheese, tartness of fermented sausage, or the color and flavor of soy sauce). In large scale industrial fermentations, microorganisms may be used to produce various enzyme preparations for many food and non-food applications, food emulsifiers and thickening agents (Chapter 9), or protein for food and feed applications (Section 2.7).

Contrary to the food spoilage situation, the optimum growth of microorganisms must be insured in food fermentation. The art and science of a successful fermentation lies in using special strains of microorganisms (*starter cultures*) selected for their particular suitability for a given task, and controlling carefully the growth conditions to achieve the maximum effect in the shortest possible time. Contamination of fermenting food by undesirable microorganisms or by compounds that inhibit culture growth may lead to flavor problems, incomplete or prolonged fermentation or, as in the case of antibiotics in milk (Chapter 6), may destroy the fermentation process altogether. Similarly, an infection of the culture by a virus (*phage*) acting as a microbial parasite may affect the fermentation process during its course. Thus, food fermentation processes, including the very important culture preparation, must be carried out under particularly stringent quality control conditions (Figure 2–5).

Culture microorganisms are supplied to food processors by several industrial companies specializing in production of carefully selected and efficient strains. The supplies are shipped either in freeze-dried form or frozen in liquid nitrogen. As mentioned in Chapter 10, these two processes will preserve viability of the microorganisms if properly controlled. The supplied cultures may be used either directly (as with frozen "redi-set" dairy cultures used in cottage cheese manufacture), or a *mother culture* will be prepared by the food processor by propaga-

FIGURE 2-5. Fermentation vessels used for manufacture of yogurt.

tion of the supplied inoculum in sterile media. The mother culture may be further transferred into "bulk culture" if the fermentation process requires a large culture inoculum. Typically, the amount of culture used in cheese-making is approximately 1–1.5 percent of the total amount of milk; for processing 100,000 kg of milk daily this amounts to 1,000–1,500 kg of bulk culture daily.

Availability of high quality cultures is one of the key requirements for production of high quality fermented food. In Switzerland, for example, where the international cheese trade is one of the major economic activites, the Swiss Federal Dairy Research Institute supplies all the approximately 1,600 cheesemakers throughout the country with cultures prepared fresh every week to insure the highest possible quality. Recent advances in food fermentation technology world wide include developments in the area of bacteriophage-resistant cultures, continuous fermentation, propagation of microorganisms as a source of food, or using microorganisms to convert food processing wastes into methane gas that can be used as a source of energy in the primary process (Figure 2–6).

FIGURE 2-6. Anaerobic fermentation of animal waste into methane for use as an energy source in a nearby cheese factory.

2.7 ROLE OF MICROORGANISMS IN HUMAN NUTRITION

Microorganisms, especially bacteria, are the most efficient biological systems for production of protein from other nutrient sources (Table 2-6). Microbial biomass contains about 50 percent crude protein. Thus, it is not surprising that industrial processes have been developed to produce protein from microorganisms grown on a suitable substrate. Engineering calculations have shown that all the protein needed by today's world population could be produced in a continuous fermentor occupying a mere 10 acres of land. This is, of course, just a theory. In practice, the protein production from single cell organisms (single cell protein or *SCP*) is much more complicated, and no large-scale industrial uses of SCP for human foods are known. Technologically, SCP production is feasible, including the harvesting of cells from the growth media, rupturing the cells, separating the cell content from the cell walls, and removing the undesirable nucleic acids from the protein extracted. The process is generally considered to be uneconomical with

Table 2-6 Biomass Production in Various Biological Systems

ORGANISM	TIME NEEDED FOR DOUBLING OF BIOMASS
Bacteria, yeast	20-120 min
Mold, algae	2-6 hrs
Plants, grass	1-2 weeks
Chicken	2-4 weeks
Hogs	4-6 weeks
Beef cattle	4-10 weeks

the present abundance of cheap food protein. A large-scale production of SCP for cattle feed has been installed recently in Europe, (Figure 2-7 on page 46) suggesting that under proper economic conditions SCP may become one of the human nutrients of the future.

The SCP process is an example of the ever increasing use of biotechnology for production of specific nutrients and other compounds important to food manufacturers. Several vitamins that can be synthesized by various microorganisms (Chapter 1) are being produced for food fortification or for direct sales to the public by specialized biochemical companies using controlled microbial fermentations. Many flavor compounds (such as diacetyl produced by several of the dairy cultures and responsible for the delicate flavor of butter made from ripened cream—Chapter 6) may be isolated by innovative industrial techniques for the needs of the modern food processing industry. Various commonly used functional ingredients (Chapter 9) such as thickening agents, ice cream stabilizers, or emulsifiers, are of microbial origin. Brewers' yeasts have been recommended traditionally as a nutritional supplement because of their high content of vitamins and other micronutrients.

These and other examples found in the contemporary food processing industry show that although microorganisms do not supply the bulk of our food nutrients, they are as important as plants and animals for the well-being of the human species.

2.8. CONTROL QUESTIONS

1. What are the differences between bacteria, yeasts and molds? What is the principal difference between bacterial and fungal spores?

FIGURE 2-7. Installation of an industrial fermentation plant for production of single-cell protein for animal feeds. (Photo courtesy of Imperial Chemical Industries PLC, Agric. Div.)

2. Many foods, when left in a household refrigerator, will spoil. What types of microorganisms may be the most prevalent causative agents of this spoilage? Why? Why do spoiled foods (e.g. milk) often taste sour?
3. Explain the terms "pathogenic organism," "food-borne illness and infection," "food poisoning," and "indicator organisms."
4. Describe the principle and aims of food processing by fermentation.
5. Laboratory procedures for enumeration of microbial populations of foods often call for mixing 1 g of a food sample with 99 ml. of sterile water and using a portion of this diluted sample for the determinative test. In a specific quality control procedure, 10 ml. of such a dilution was distributed for incubation in three different petri dishes with nutrient agar. What would be the microbial load of your food sample if the three incubated petri dishes showed 520, 478 and 685 CFU/dish?
6. What is "single-cell-protein" (SPC), and why is it not used for food production at the present time?

Suggested reference books —
Food Microbiology, Enumeration of Microorganisms, Water Activity

Banwart, G. J. 1979. *Basic Food Microbiology*. Westport, Conn.: AVI Publ. Co.

Bradley, H. and Sundberg, C. 1975. *Keeping Food Safe*. Garden City: Doubleday Co.

Duckworth, R. B. 1975. *Water Relations of Foods*. London: Acad. Press.

Fields, M. L. 1979. *Fundamentals of Food Microbiology*. Westport, Conn.: AVI Publ. Co.

Frazier, W. C. and Westhoff, D. C. 1978. *Food Microbiology*, New York, N.Y.: McGraw-Hill.

Riemann, H., and Bryan, F. L. 1979. *Food-borne Infections and Intoxications*. New York: Acad. Press.

Roberts, H. R. 1981. *Food Safety*. New York: John Wiley & Sons.

Roberts, T. A., Hobbs, G., Christian, H. J. B., and Skovgaard, N. 1981. *Psychotrophic Microorganisms in Spoilage and Pathogenicity*. London: Acad. Press.

Roberts, T. A. and Skinner, F. A. 1983. *Food Microbiology— Advances and Prospects*. London: Acad. Press.

Thatcher, F. S. and Clark, D. S. 1973. *Microorganisms in Foods—their significance and methods of enumeration*. Toronto, Ont.: University of Toronto Press.

Troller, J. A. and Christian, J. H. B. 1978. *Water Activity and Food*. New York: Acad. Press.

3

The Food Industry

3.1 FOOD MANUFACTURING AND PRESERVATION

Most of the agricultural products used as human food cannot be used directly without some processing. Even fresh produce such as apples, potatoes, lettuce, or cauliflower must be washed, trimmed, sorted, or packaged. More often, a rather extensive processing is necessary to convert primary agricultural products such as beef cattle, sunflower seeds, wheat kernels, sugar beet, or milk, to meat, cooking oil, flour and bread, table sugar, and butter or cheese. Roughly speaking, most of the activities of the primary food processing industry are concerned with *manufacturing* foods and raw food materials from agricultural commodities. In the above examples, flour, cooking oil, or granulated sugar can be considered raw materials for further food manufacturing by the *secondary* food processing industries such as bakeries, cheese factories, wineries, breweries, and many other industrial producers of the foods found on supermarket shelves. Some of the secondary food processors' roles are still predominantly manufacturing, since they are producing food ready for human consumption (such as bread) from raw materials which are not particularly suitable for food (flour, water, yeast). Other secondary food manufacturing processes combine a *manufacturing* function with an element of *preservation*. Numerous food products (such as pickled cucumbers, sauerkraut, canned ham, frozen orange juice, dried milk and many others) are not only industrially produced food items, but they also have a much longer shelf–life than the original agricultural commodities they are made from. This preservation function is even more enhanced in the products of *tertiary* food processors who transform primary or secondary food materials into shelf-stable convenience foods such as canned stew, instant coffee, ice cream–filled desserts, or complete, ready-to-serve frozen meals.

The food processing industry has a two-fold role: (1) to manufacture edible food items from mostly inedible agricultural products, and (2) to preserve the oversupply of agricultural products available at the time of harvest for consumption later in the year. These two sides of food processing are often closely interrelated. For example, cheese is made both to utilize the oversupply of milk available at certain times of the year and to produce a different food for which a market exists.

The contemporary, sophisticated food processing industry has developed from the original small-scale, village-type or home-based operations. One of the important reasons for this development is the continuing industrialization of Western societies, resulting in virtual separation of food producers—the farmers—from the large food markets in big cities. As shown in Figure 3–1, the food chain of today is different from the simple farmer-customer relationships of yesterday. Several middle-men must be involved in the food processing and distribution system today. Delivering food produced on large North American farms to the hungry domestic or foreign markets is simply too big a task for a farmer or a small entrepreneur, especially when one farm

```
         Yesterday                          Today

   ┌───────────────────┐            ┌───────────────────┐
   │ Farmer = Processor │           │       Farmer       │
   └───────────────────┘            └───────────────────┘
             │                                │
             ▼                                ▼
   ┌───────────────────┐              Raw material delivery
   │ Retailer and/or consumer │               │
   └───────────────────┘                      ▼
       Immediate family                   Processor
         Neighbors                            │
        Village level                         ▼
                                          Distributor
                                              │
                                              ▼
                                           Retailer
                                              │
                                              ▼
                                    ┌───────────────────┐
                                    │     Consumer      │
                                    └───────────────────┘
                                     Across town, across country
                                         or across the world
```

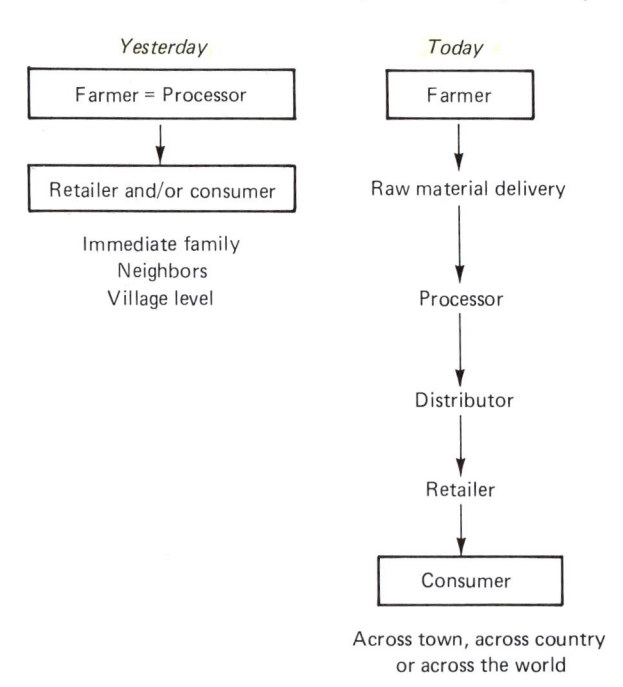

FIGURE 3–1. The food chains of yesterday and today.

Table 3-1 Agricultural Population and Food Costs in Selected Countries[a]

COUNTRY	TOTAL POPULATION (MILLIONS)	POPULATION ACTIVE IN PRIMARY AGRICULTURE (%) TOTAL	RATIO OF TOTAL POPULATION TO FARM POPULATION	DISPOSABLE INCOME ($/PERSON)	FRACTION OF INCOME SPENT ON FOOD[c] (%)
Australia	14.6	6.0	15	9,348	n.a.
Canada	23.9	5.3	18	9,133	15.4
Germany	61.5	4.3	22	11,759	22.3
Japan	116.8	11.8	8	7,672	20.2
Nigeria	74.1	57.2	2	510	n.a.
USSR	265.5	17.3	6	n.a.[b]	n.a.
USA	227.7	2.3	44	10,094	15.5

[a] Source: United Nations Statistical Yearbook, 1983. Adaptation Courtesy of Prof J. Richter, Ag. Econ. Dept., University of Alberta.
[b] Not available.
[c] Source: *National Accounts of OECD Countries*, Vol II, (Paris: 1982, and U.N. Statistical Yearbook, 1980).

worker in the U.S. has to produce enough food to feed forty-four people (Table 3–1).

Naturally, all the participants in the food chain from the farmers to the processors, distributors, transportation companies, or retailers are business-oriented, just as car manufacturers or movie-makers are. With profits being the main driving force of any independent business activity, it is inevitable that the complexity and sophistication of the food chain will have an effect on retail food prices. Without the mechanization and concentration of food processing into large integrated systems, however, food would be scarce and highly seasonal, and its cost would be much higher than the present 16 percent of the average North American family income. In countries with less sophisticated food processing and distribution systems food bills are often consuming most of the family earnings.

3.2 FOOD PROCESSING AS AN APPLIED SCIENCE

Historically, food manufacturing operations developed largely from tradition, incidental observations, or primitive craftsmanship. When people discovered that seeds of certain grasses—specifically wheat—could be ground to produce flour, which is useful for making bread, they created a food processing system. It was only much later that we learned why wheat is suitable for bread-baking but barley is not. This knowledge then helped us to advance rapidly in the science and technology of baking.

Today's food industry is based on at least six different areas of scientific endeavors, not including the agricultural sciences. The realm of food science (of which food processing is the practical application) is usually taken as "everything that happens beyond the farmer's gate." The following fields of study are represented in the structure or personnel of a modern industrial food processing chain:

Engineering. Processing machines, factories, transportation systems, computer controls and other technological equipment must be designed and operated efficiently in order to achieve the desired transformation of agricultural commodities into food.

Marketing. From the broadest viewpoint, including development, testing, and quality improvement of food products to suit customers' needs, this is the science of creating and providing food which is acceptable, palatable, desirable, and therefore can be marketed successfully.

Economics. In the free enterprise system, all food processing operations including all ancillary functions such as utilities, administration, raw materials handling, distribution, or quality control, must be properly financed and must ultimately result in a profitable operation, otherwise the enterprise could not exist.

Chemistry. Knowledge and facilities must be available for the necessary food analyses, to prevent undesirable chemical reactions leading to spoilage in final products, or to accomplish desirable chemical reactions in food manufacture such as coagulation of milk to produce cheese.

Microbiology. The three major concerns of food microbiologists are the prevention of food spoilage, elimination of harmful bacteria causing food-borne illnesses, or maximization of desirable microbial actions such as in fermentation of yogurt, beer, or wine.

Nutrition. All food processing activities must result in providing adequate amounts of necessary nutrients while eliminating substances that are deleterious to human health or have anti-nutritional properties.

In the modern food processing operations, all of the above aspects are interrelated. Food manufacturing plants are designed by engineers for the most economical operation, minimum hazards of microbial contamination, and maximum nutrient retention. Traditional and new food products are being manufactured to be more attractive, nutritious or cheaper than the competitors' products to attract a larger share of the market. Processes are being developed to utilize new knowledge of biochemical or enzymatic reactions resulting in lower energy consumption, better utilization of raw materials, or less waste water. In large industrial enterprises, specialized knowledge from other fields of science or art is also required. Thus, compliance with food laws and regulations must be assured by lawyers; design of food packaging labels is often the work of an artist; the well-being of workers may be under the control of a psychologist. In short, a successful food processor must pay attention to all the facets of the diverse field of food science and many other related areas.

3.3 INDUSTRIAL PLANT ORGANIZATION

While it is obvious that manufacturing of food products is the primary function of an industrial food plant, there are many areas of technical or human input into this process. Long gone are days when the farmer was processing and marketing his own products and was his own office manager, plant engineer, quality control officer, marketing manager, and accountant. In the modern food processing factory, there are at least four distinctly different areas of employee activity common to most industrial plants.

Technology.

The heart of a manufacturing plant is the process technology. The combination of steps necessary to produce a food item is referred to as a *processing line*. It consists of various pieces of machinery designed to provide specific required treatments to the product as it moves along the line. As an example, a schematic diagram of a processing line for production of dried milk powder and butter from fresh liquid milk is shown in Figure 3-2. Sometimes a rather sophisticated computer control center is provided to operate the machines automatically; other lines or individual machines are operated manually by the employees. The design of a processing line is based on technological requirements for the given process, which in turn is influenced by the various chemical, physical, microbiological, or engineering properties of the raw materials and quality requirements for the finished food product. A description and exploration of the main technological steps in manufacturing our most common foods constitutes the bulk of this book.

Support Services.

In the above example of a dairy processing plant (Figure 3-2), the final products could not be manufactured by machines alone. *Electricity* is

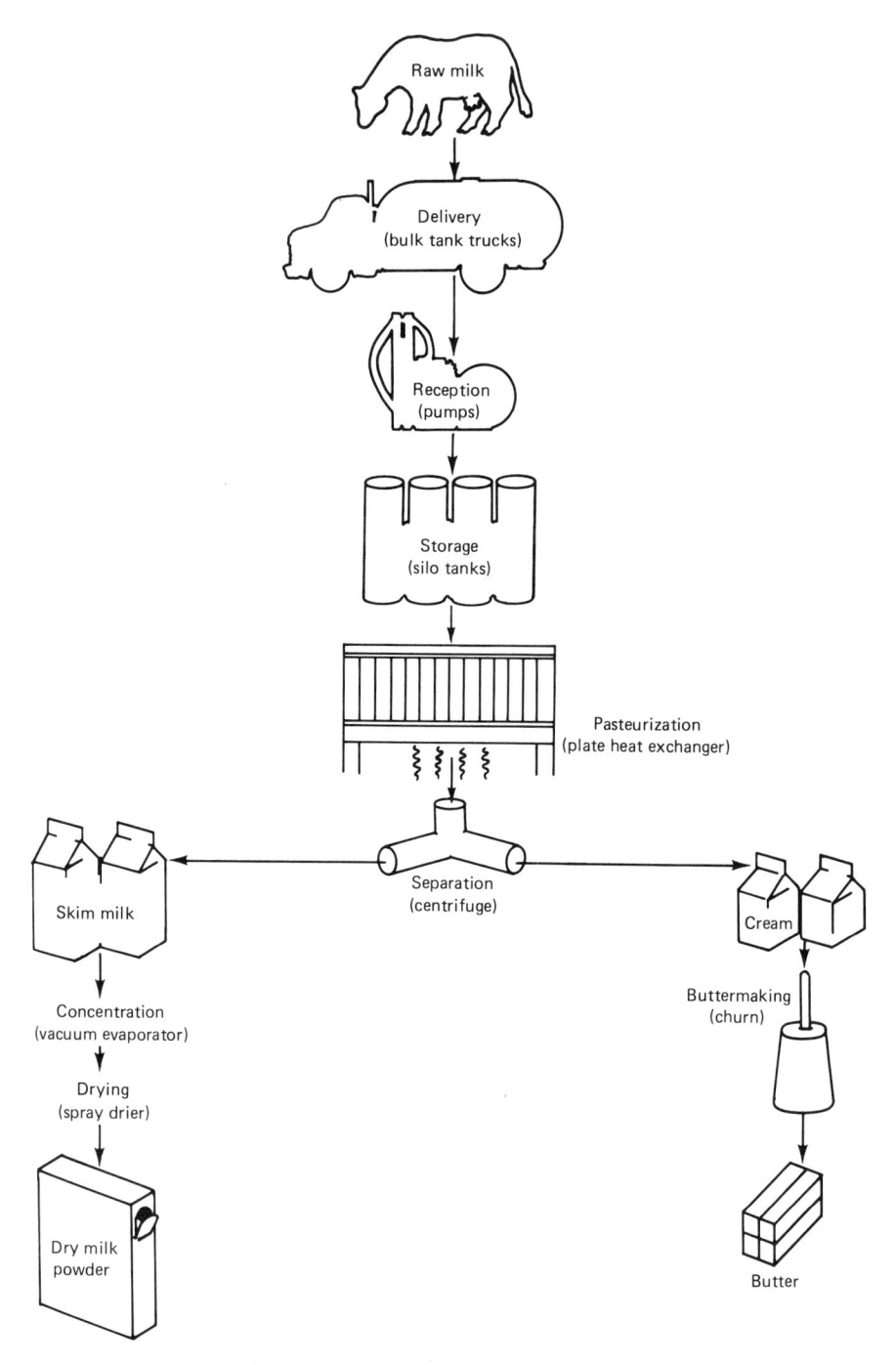

FIGURE 3-2. A schematic diagram of a dairy processing line for the manufacture of butter and non-fat dry milk.

needed for operation of the machinery (pumps, centrifuges, butter-churning equipment) and their control systems. *Water* must be available to cool down the milk in a plate heat exchanger (see Chapter 6), to wash the machinery every day, and to provide the *steam* needed for general heating purposes and in evaporation of the milk for drying. The steam is generated in a boiler, where *fuel* (coal, oil, or natural gas) must be burned to provide the necessary heat. Storage of the finished food product, as well as some of the processing steps, require *refrigeration* (see Chapter 10). In automated dairies and other food plants compressed *air* is used to control material flow by air-activated valves. In the manufacture of dry milk a large amount of hot air is also needed for the spray-drying process (Chapter 10).

Environmental pollution from processing wastes, wash water, spills of milk, and discarded whey (if the dairy plant is also processing milk into cheese) must be avoided by an efficient *waste-treatment facility*. Untreated waste waters from many food processing plants have relatively high content of biological matter which, if discarded into rivers and other water bodies, would serve as a nutrient supply for various microorganisms. The microbial growth would, in turn, deplete the water of oxygen and this would result in severe damage to aquatic life. The standard test for strength of waste waters is based on determining the amount of oxygen that would be used by bacteria growing on the waste for five days (this is termed the biological oxygen demand or BOD_5).

Some of the above services and utilities, especially the supply of water, electricity and sometimes waste treatment, are provided by municipalities at cost. In other cases such as refrigeration, steam or compressed air, the factory must have its own resources with the necessary manpower, machinery and supplies. Often, a standby electricity generating capacity is desirable to avoid unexpected food quality problems that would result from a main supply failure. The cost of support machinery and services needed for food processing can contribute substantially to the total cost of the final product.

Some of the important design decisions regarding plant location or selection of alternative technology may be based on availability and local costs of services. Continuing with our example of the dairy processing plant, the selection of the drying technology may depend on the local availability, economy, and energy content of the alternative fuel sources needed to heat the drying air. Table 3–2 gives some orientation energy values for the main industrial fuels. The location of a cheese plant may be decided by the costs and availability of municipal waste treatment, as waste treatment is usually a very expensive and unproductive component of the food manufacturing process. Principles of the main alternative waste treatment techniques available to food processors are summarized in Table 3–3.

56   The Food Industry

Table 3-2   Approximate Energy Content of Selected Industrial Fuels

MATERIAL	ENERGY CONTENT [a] (MJ/kg)
Coal (average value)	30
Oil	45
Natural gas	37

[a] 1 MJ = 1000 kJ. For definition of kJ see Table 1-7.

Operation and control of some of these support services may be a well-defined function requiring the knowledge of a trained specialist engineer. Food processors, however, must be familiar at least with the principles and the potential problems that may arise in the support services area.

Storage, Warehousing and Distribution.

All raw materials such as dug-up potatoes, live cattle, or freshly collected milk, must be stored in sufficient quantities to provide a steady supply for the manufacturing process. Storage conditions are often carefully controlled to avoid deterioration in quality; nobody can manufacture high quality food products from poor quality raw material. In some cases, a suitable storage regimen is an integral part of the technological process; as an example, reconditioning of potatoes is discussed in Chapter 4. The organization and control of a raw material supply and

Table 3-3   Principles of Different Waste Water Treatment Processes Used By the Food Industry

TREATMENT TECHNIQUE	PRINCIPLE OF OPERATION
Septic Tank	Settling and uncontrolled anaerobic digestion in an enclosed earthern tank; must be periodically emptied.
Settling Lagoon	Removal of suspended solids by gravity sedimentation.
Activated Sludge, Aerobic Pond	Digestion of dissolved organic matter by aerobic bacteria, oxygen supplied by active aeration; loose bacteria form sludge.
Trickling Filter	Aerobic digestion by bacteria attached on porous matters; waste water trickled through the "filter".
Anaerobic Filter, Methane Fermentor	Digestion of organic matter in complete absence of oxygen; anaerobic bacteria attached on filter support or loose in the fermentor.

storage system is an important and complex task facing most food processors, especially when the supply is being received daily such as in our earlier example of dairy processing.

Warehousing and distribution of final products is another difficult and often costly aspect of the food processing industry. Refrigerated storage areas, transportation vehicles, material handling equipment such as fork-lift trucks, palletizing machinery for easier load controls, and other sophisticated systems occupy a large proportion of the factory's manpower and resources (Figure 3-3). Inventory control and minimization of pilferage is often a difficult management problem. All supplementary materials and supplies required for the primary function, the manufacturing of a food product, must be also available as needed. Efficiency in ordering and storing of all the necessary packaging materials, cleaning supplies, secondary ingredients, spices, food additives, supplementary nutrient preparations, fuels, even office and laboratory supplies, complements a successful operation of the plant. On the contrary, poor quality of unreliable supplies or insufficient warehousing systems may cause great difficulties in the manufacturing process.

FIGURE 3-3. Automated product palletization and handling systems in a large dairy factory.

Office and Laboratory Services.

All the above mentioned systems could not function without effective management and control. The desired food products could not be manufactured without a steady supply of the necessary raw materials which have to be contracted for and purchased from various suppliers. The final products could not be sold without a proper marketing network, including advertising and delivery systems. The company staff must be hired, supervised and paid. Product quality must be constantly monitored by a quality control laboratory in order to satisfy the various legal requirements, the manufacturer's own quality criteria, or the specifications of other industrial companies using the finished products. Development of new food items for the retail market is important for many companies in order to keep ahead of competitors.

While including all the above functions, the actual organization of a food processing company may vary greatly. A generalized organizational chart applicable to many food companies is shown in Figure 3–4. Depending on the size of the company, auxiliary departments may constitute an important component in the firm's total structure. Some of the large diversified food processors may have several hundred scientists and support staff working just in the research-development department alone! On the contrary, in some of the smallest food companies, all the above functions will be looked after by only a handful of

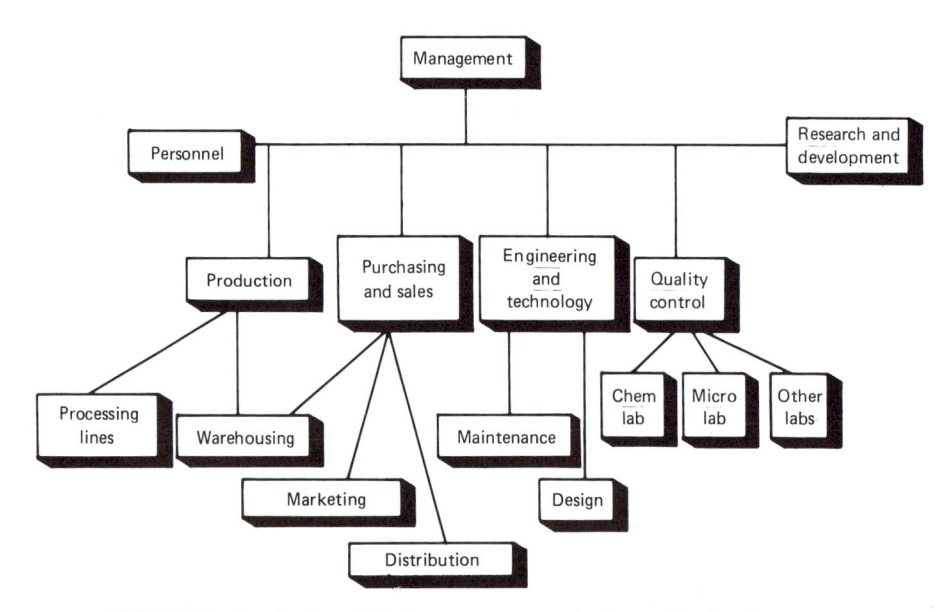

FIGURE 3–4. A simplified organizational chart of a food company.

employees often fulfilling several of the tasks at the same time. Less essential activities, such as new product development, may be contracted out to independent consultants or private laboratories, carried out in government-owned pilot plant testing facilities, or may not be covered at all.

3.4 PRODUCT DEVELOPMENT AND QUALITY CONTROL

Of all the functions discussed, product development research and quality control require the most diversified knowledge of the food science field. Typically, these positions are held by graduates of university food science programs. The philosophy, goals, and methods used to accomplish these two functions of food processing are fundamentally different. The research and development efforts are aimed at *new, nonexisting* or *improved* products or processes, methodologies, or laboratory procedures. Thus, innovative approaches and nonroutine, independent experimentation are often required. In large food processing companies, basic research in biochemistry, microbiology, enzymology, chemistry or engineering may be encouraged as needed for development of new foods or new processes. Although there are similarities in the work of many research and development departments of various companies, it would be difficult to suggest a generalized pattern for food research and development activities.

On the contrary, the quality control (or quality assurance) procedures methodically follow a fairly standard set of routines since quality assurance is concerned with *existing* products and processes. The major function of the Quality Control (QC) department is to monitor continuously the required parameters of incoming raw materials and finished food products. To achieve this objective, the QC efforts may be concerned with all the major areas of food processing as discussed in sections 3.2 and 3.3. Thus, QC personnel must insure compliance with the various compositional standards of identity (usually proximate composition) of the various products manufactured, as well as compliance with the various government regulations concerning the use of various additives. In the microbiological laboratory, constant monitoring of microbial content of the ingredients and final products is carried out as discussed in Chapter 2.

The sensory qualities of final products, raw material ingredients and competitors' products should also be routinely checked. A complete sensory evaluation laboratory should include facilities for *taste panel* work (Figure 3–5) where experienced and discriminating tasters are used and their results are statistically evaluated. Some sensory attributes of food products (especially color or texture) can be mea-

FIGURE 3-5. A laboratory for sensory evaluation of food products by taste panels.

sured by instruments (Figure 3-6) and correlated with quality aspects important to consumers. Evaluation of flavors, odors and tastes usually requires repeated judgements by panels of human subjects since individual flavor perceptions and likings may differ substantially. Proper selection and management of taste panel procedures is one of the most tedious and often abused components of quality control work. Table 3-4 summarizes the major functions of a complete quality control department, while typical routine quality control procedures are listed in Table 3-5.

Although philosophically research and development is different from quality control, these two functions are sometimes combined in one department or even one person, especially in smaller companies. The QC manager should have the power to halt production if the appropriate quality criteria are not maintained; however, this is sometimes in direct contrast with the objectives of the production manager who is responsible for the manufacture of products required to fill customer orders. One of the important tasks of both the production and quality control managers is to insure that all final products comply with food laws, standards and other regulations issued by the various government agencies controlling the food processing activities.

FIGURE 3-6. The Instron texture-measuring apparatus for instrumental evaluation of food structure and consistency. (Photo courtesy of the Instron Corporation.)

Table 3-4 Major Functions of Quality Control in the Food Industry

TYPE OF ACTIVITY	EXAMPLE
1. Compliance with Specifications	Legal requirements, internal company standards, shelf-life tests, buyer's specifications, establishment of own specifications.
2. Test Procedures	Develop and perform tests for raw materials, finished products, and in-process tests.
3. Sampling Schedules	Select or develop a suitable sampling schedule to maximize the probability of detection while minimizing the amount of work needed (MIL STD 105; Shewhart charts; statistical procedures).
4. Records and Reporting	Maintenance of all quality control records so that consumer complaints or legal problems can be dealt with. Develop appropriate forms, reporting mechanisms, etc.
5. Trouble Shooting	Often in cooperation with R & D - solve various problems caused by poor quality raw materials (modification of a technological process), erratic supplies, malfunctioning machines, investigate reasons for poor quality final product to avoid repetition.
6. Special Problems	Bad lots, consumer complaints, production problems, personnel training, short courses.

3.5 GOVERNMENT REGULATIONS

The food industry as a whole is one of the most regulated free enterprise activities in North America. In the U.S. on the federal level alone there are no less than six agencies that have regulatory responsibilities over some aspect of food processing, distribution and merchandizing. Similar fragmentation of responsibilities exists in Canada. Table 3-6 lists the major regulatory agencies in Canada and the U.S. and briefly explains their roles.

In both countries the dominant aspect of governmental involvement in the food industry is to protect the consumer against fraudulent and unacceptable practices that may have been possible with past food supply methods, and to assure food safety.

Government regulatory powers have been formulated in acts of the U.S. Congress (1938) and the Canadian Parliament (1953) and in several later amendments to these acts. The agencies with primary responsibilities for enforcement of the legislation are the Food and Drug Administration (FDA) in the U.S., and the Food Directorate of the Health Protection Branch (HPB) in Canada. In both countries, the enforcement

Table 3–5 Examples of Routine Quality Control Procedures Used in the Food Industry

CATEGORY OF TESTS	SPECIFIC EXAMPLES AND PURPOSE
Chemical	Proximate analyses of raw materials and finished products (protein, fat, moisture)
	Analyses for special nutrients (calcium, vitamins)
	Analyses for presence of antibiotics (raw milk)
	Analyses of waste waters
	Analyses of cleaning and sanitation compounds
Microbiological	Plate counts of raw materials and finished products (total count, coliforms, yeasts and molds, psychrotrophs, other specific tests)
	Testing new cultures (fermented products)
	Sampling from processing lines for detection of microbiological hazards and problems
	Determination of efficiency of cleaning procedures (swab testing of equipment surfaces)
	Identification of microorganisms in case of problems
Sensory	Sensory tests for grading of products (cheese, butter)
	Routine testing of own products for maintenance of uniform quality
	Periodic tasting of own products against same products of a competitor
	Instrumental testing (color, texture, aroma profile)
Other	Accelerated shelf-life testing of finished products
	Random control of temperature in display cabinets of retail stores

agencies are included in the structure of federal departments dealing with citizens' health (Department of Health and Human Services, formerly Health, Education and Welfare in the U.S., and Health and Welfare in Canada). The general approach of existing regulations is restrictive in that the regulations define what may and what may not constitute—or be used in—a given food item. In some cases the regulations have less to do with protection of consumer safety than with assurance of quality through component or process specification. Some of these regulations may be referred to as standards of identity since they specify how a food product must be made. These regulatory standards protect the consumer against the possibility of deceptive practice by an unscrupulous manufacturer and insure that the product contains the proper amount of given nutrients.

Both in Canada and the U.S. the main government regulatory activities concerning food safety (and often even food manufacturing

Table 3–6 Federal Government Agencies With Regulatory Responsibilities Related to Food Processing

BRANCH OF GOVERNMENT AND AGENCY		RESPONSIBILITY
U.S.A.	Canada	
Dept. of Health and Human Services (U.S. Public Health Service - USPHS) (Food and Drug Administration - FDA)	Health and Welfare Canada (Health Protection Branch - HPB)	Enforcement of food legislation regarding food wholesomeness and aspects concerning human health and food process sanitation; food served by transportation companies; advisory capacity in milk, food and shellfish sanitation.
Department of Agriculture (USDA)	Agriculture Canada (Inspection Branch)	Enforcement of standards of identity, plant sanitation inspection, veterinary inspection of meat, grading of meat, dairy, fruit and vegetable products
Federal Trade Commission (FTC)	Consumer and Corporate Affairs Canada	Enforcement of all food legislation related to packaging, labeling, and advertising of food products
Environmental Protection Agency (EPA)	Environment Canada	Control and enforcement of food processes resulting in pollution of waters, land or air; control of pesticide application; control of water supplies
Department of Commerce (National Marines and Fisheries Bureau)	Fisheries Canada	Regulation of catch, imports and processing of fish products
Department of Treasury (Bureau of Alcohol, Tobacco and Firearms - BATF)	Provincial Liquor Control Boards	Production and sales of all alcoholic beverages
Department of Commerce and Dept. of Agriculture (USDA)	Industry, Trade and Commerce (ITC), and Agriculture Canada	Regulatory powers of international food trade

processes) are removed from the jurisdiction of the Agriculture Departments (USDA or Agriculture Canada) which oversee the actual production and processing of most foods. The Agriculture Departments, however, also participate in government regulatory activities. In both countries the safety of meat and poultry products is assured by veterinary inspectors of the USDA or Agriculture Canada. Personnel from these departments are also responsible for enforcement of hygienic practices and sanitary food processing operations, assigning quality grades to meat, cheese, and fresh produce crossing interstate or inter-provincial boundaries, and controlling compositional standards of finished products.

Many other government agencies are involved in the regulation of food processors. As shown in Table 3–6, enforcement of labeling and packaging is the responsibility of the Federal Trade Commission in the U.S. and the Department of Consumer & Corporate Affairs in Canada. Wine, beer, and liquor products are not controlled in the same manner as fishing and fishery products. There are also provincial control agencies, municipal boards of health, agencies responsible for enforcement of environmental controls, agencies protecting workers' safety and labor practices and insuring competitiveness of the industrial firms. All these regulatory activities can be justified on their own merits; for the processors, they represent an increasingly burdensome, costly, and often counterproductive interference in the sphere of their main industrial activity.

3.6 MATERIAL AND ENERGY BALANCES

Most processed foods are manufactured by mixing or separating several components or materials. Bread is made from flour, water, shortening, yeast, and specialty ingredients which must be mixed in appropriate proportions; sausage is made from various meat cuts, seasonings, and flours; whiskey may be blended from various traditional products; orange juice concentrate is produced by evaporating water from the fresh juice to a desired total solids composition. The mixtures could vary widely in composition or functional properties if inappropriate amounts of individual materials were used. Thus, it is necessary to know how to calculate the exact amounts of required ingredients that must be used for any particular product.

To prepare a given amount of a product (a "batch" in industrial jargon), the food processor must determine how much of each ingredient he needs to satisfy the various requirements of legal standards, quality limits, or flavor and color intensity. Some of the most important routine calculations performed by the food processor every day deal with formulations of the various products to be made.

The basis for these calculations (known to engineers as *material* or *mass balances*) is the law of the conservation of mass, which says that *"mass can be neither created nor destroyed"*. Thus, all mass (measured usually as weight) of the individual components entering the process must be accounted for in the final products, by-products, wastes, or product losses. Knowledge of proximate composition of the individual components, the desired quantity of the final product, and the understanding of the processing sequence are required for solving formulation problems. The calculations take the form of a system of linear equations in as many unknowns as we have components, and the mathematical procedures are straightforward once the equations are formulated. The step-by-step mass balance calculation process is outlined in detail in Appendix A.

Mass balance determinations are also useful for control of process effectiveness. In our previous example (Figure 3–2), the production losses of dry milk powder can be accounted for accurately by knowing how much liquid milk we processed, what was the total solids composition of the raw material and finished product, and how much finished product we produced. This simple inventory control calculation is shown in summary in Table 3–7. In a similar fashion, energy consumption in a food processing plant can be controlled by procedures known as *energy balance* computations. Just as in the mass balance approach, energy balances are based on the law of the conservation of energy. However, since energy can exist in various forms, this must be taken into account in the calculations. Energy balances are particularly useful in determining the needs of steam or hot water for heating, or the needs of refrigeration capacity for energy removal during cooling (Chapter 10). An example of a simple energy balance is included in Appendix A.

3.7 SOLUBILITY AND CONCENTRATION IN LIQUID FOODS

Many liquid materials used or produced by the food industry are solutions of one or more solutes in a solvent, most often water. Fruit juices, milk, beer and wines, pickling brines and many other food products and ingredients can be best characterized by their *solute concentration*, denoting the amount of the particular material dissolved in water. The concentration, however, may be expressed in many ways. For the mass and energy balance calculations it is important to pay attention to the basis used for the concentration data.

In chemistry, concentration of solutions is usually given in molarity, molality, mole fraction, or normality. For liquid food systems,

Table 3-7 Example of Mass Balance Calculation for Inventory Control in Processing of Spray-Dried Milk

INFORMATION SEQUENCE	DATA OBTAINED	DATA CALCULATED
I. Raw Material		
1. Amount of skim milk processed	10,000 kg	—
2. Total solids content	9.0%	—
3. Amount of total solids available (line 1 × line 2)	—	900 kg
II. Finished Product		
4. Number of bags with skim milk powder produced	18	—
5. Nominal weight of product per bag	50 kg	—
6. Amount of dry milk powder produced	—	900 kg
7. Average moisture content of dry powder (by analysis)	3%	—
8. Total moisture in product (line 6 × line 7)	—	27 kg
9. Total amount of dry skim milk solids (line 6 minus line 8)	—	873 kg
10. Product loss percentage ($\frac{\text{line 6} - \text{line 9}}{\text{line 9}} \times 100\%$)	—	3%

which are often mixtures of many components, the chemical approach would be impractical. Instead, the typical unit used by the food industry is the *percent concentration.* In most cases, the basis is the *total weight* of the *solution*, and the concentration is expressed as % weight/weight (w/w). Thus, a 20 percent (w/w) sucrose solution describes a system in which 20 parts of sucrose (by weight, for example 20 kg) are dissolved in 80 parts by weight (thus, 80 kg) of water. Of the total weight of the system (100 kg), 20 percent is attributable to sucrose. This is identical to the proximate composition explained in Chapter 1.

Some liquid foods, which contain two liquids mixed by volume, may be characterized by a % volume/volume (v/v) concentration. Typically, the alcohol content of beer or wine is expressed in v/v percentages; a 12 percent (v/v) alcohol content of a dry white wine means that there are 12 volumetric parts (e.g. 120 mL) of alcohol in 100 parts (1,000 mL) of the wine. The w/w and v/v expressions are not interchangeable because of the differences in density of the individual components. This can be best illustrated by using the above mentioned wine as an example.

The density of alcohol is approximately 0.8 g/mL; the alcohol content by weight in 1,000 mL of wine can be calculated as

$$0.8 \times 120 = 96 \text{ g} \qquad \text{(Equation 3-1)}$$

Although there are some other minor solutes besides alcohol present, their effect is negligible in this example, and we can consider the remaining 880 ml of wine as water. Thus, since the density of water is 1 g/mL, the total mass of this system is

$$96 + 880 = 976 \text{ g} \qquad \text{(Equation 3-2)}$$

and the w/w alcohol concentration is 96 g (Equation 3-1) in 976 g (Equation 3-2), or

$$\frac{96}{976} = 9.84\% \text{ (w/w)} \qquad \text{(Equation 3-3)}$$

Several other expressions of concentration are traditionally used by some food industries, or in food research literature. Table 3-8 gives a list with a brief explanation of the more commonly used terms. It is imperative that the concentration basis is clearly understood when dealing with mass balance or formulation problems.

The most frequently encountered food solutes, sugars and salts, have a limited *solubility* in water; only a certain definite amount of each solute can be dissolved in a given amount of water. When the solubility or *saturation value* is exceeded, the excess solute will crystallize or precipitate. Determining the concentration and saturation values for the various food materials is one of the important tasks facing food processors, especially sugar manufacturers, dairy processors, and companies processing fruits and vegetables. Concentration of alcohol in alcoholic beverages as well as concentration of sugars in various fruit juices, jams and other fruit and vegetable products must conform to the legal specifications or label declarations.

3.8 CONTROL QUESTIONS

1. Preparation of a festive roast beef dinner at home is akin to industrial food processing. Show how the six main areas of scientific endeavors, which constitute the realm of food processing, are represented in the preparation and serving of the roast beef.
2. What is a food processing line and what would be some of the technological operations on which a simple line for producing packaged apple juice from fresh apples would be based?
3. Why is waste water treatment an important component of most

Table 3-8 Alternatives for Expressing Concentrations of Food Solutions

EXPRESSION BASIS	SYMBOL	INTERPRETATION	EXAMPLE
Weight/weight	w/w	Weight of solute per total weight of solution	10 kg sugar/100 kg sugar solution = 10% w/w
Volume/volume	v/v	Volume of solute per total volume of solution	12 mL alcohol/100 mL wine = 12% v/v
Weight/volume	w/v	Weight of solute per total volume of solution	5 g NaCl/100 mL salt solution
Weight/weight of solvent	w/w_s	Weight of solute per standard weight of pure solvent	10 kg sugar/100 kg water
Degree Brix	°Brix	Industry term for concentration of sugar solutions; equivalent to % w/w	15°Brix = 15% w/w sugar solution
Degree salometer	°S	Industry term for concentration of salt brine; given in % of total saturation which is approx. 25 g salt/100 g solution, thus 4°S = 1% w/w	20°S = 5% w/w salt solution
Degree Baumé	°Bé	Empirical concentration scale based on specific gravity of individual solutions (need appropriate tables)	5°Bé (identify solution)
Degree Lactometer	°L	Similar to Bé scale, specific for dairy products	4°L (identify skim or whole milk)

food processing plants? Explain the principal test used for determining the strength of the waste waters.

4. A can of frozen concentrated orange juice bought in a grocery store was found to contain 200 g of the frozen juice. The total solids content of the juice was found to be 32.5%. To make a drink, the directions on the can call for adding three cans of water. If we say that a can of water adds approx. 150 mL of water, what will be the concentration in % total solids of the diluted juice?

5. In preparation of the concentrated orange juice from above, the freshly squeezed juice is evaporated to remove a certain amount of water. If the fresh juice contains 8% total solids, calculate the final product concentration (after evaporation) if we started with 50,000 kg of the fresh juice and obtained 16,000 kg of the concentrate. Express the result in w/w and w/w solvent!

6. What is the legal basis for government control of the food processing companies and why is composition of some food products legally defined?

7. How much salt and water do you have to mix together to prepare a 60°S (Salometer) brine?

Suggested reference books -
Food Industry Organization, Regulation and Quality Control

Grant, E. L. 1964. *Statistical Quality Control.* New York: McGraw-Hill.
Greig, W. S. 1971. *The Economics of Food Processing.* Westport, Conn.: AVI Publ. Co.
Health Protection Branch. 1979. *Health Protection and Food Laws.* Health and Welfare Canada, Ottawa: Educational Services.
Hui, Y. H. 1979. *United States Food Laws, Regulations and Standards.* Somerset, N.J.: John Wiley & Sons.
Kramer, A., and Twigg, B. A. 1973. *Quality Control for the Food Industry.* Westport, Conn.: AVI Publ. Co.
Larmond, E. 1977. *Laboratory Methods for Sensory Evaluation of Food.* Publ. 1637. Agriculture Canada, Ottawa: Research Branch
Potter, N. N. 1978. *Food Science.* Westport, Conn.: AVI Publ. Co.
Schultz, H. N. 1981. *Food Law Handbook.* Westport, Conn.: AVI Publ. Co.
Whelan, E. M. and Stare, F. J. 1975. *Panic in the Pantry.* New York: Atheneum.

4

Processing of Fruits and Vegetables

4.1 CHARACTERISTICS OF FRESH FRUITS AND VEGETABLES

The main role of fruits and vegetables in human nutrition is to provide certain micronutrients, particularly vitamins C and A, some trace minerals and some fiber, and to contribute to the pleasure of eating. Because these popular agricultural products are consumed or processed fresh, they are high in moisture (Table 1–3) and consequently, their food energy content is low. Intake of other foods supplying our daily allotment of required macronutrients is often enhanced by fruits and vegetables, or products made from them.

Fruits and vegetables are foods of plant origin. From the botanical viewpoint, many vegetables—such as green peppers, cucumbers, tomatoes, green beans, etc.—are in fact fruits, i.e. those parts of plants that develop from fertilized blossoms and contain seeds. Other parts of certain plants—roots, stalks, blossoms, leaves, or the whole plant—may be used as a "vegetable," but rarely as a "fruit." Government food regulations do not provide any definition of what may be used as a fruit or a vegetable; the accepted distinction seems to be based largely on the traditional consumer usage of these plant food materials. Table 4–1 illustrates the botanical differences of the most common vegetables and gives some typical compositional data.

Several common characteristics of fruits and vegetables, typical for agricultural commodities of plant origin, are important to the fruit and vegetable processor. *Seasonality* of production results in highly imbalanced supply of a perishable raw material in a relatively short time

Table 4-1 Botanical and Compositional[a] Characteristics of Selected Vegetables (Raw, Edible Parts)

VEGETABLE	PLANT PART	H_2O	PROTEIN	CARBOHY-DRATE
		——— % WEIGHT ———
Carrots	Root	88.2	1.0	8.6
Potato	Tuber	79.8	2.1	17.1
Celery	Stalk	94.1	0.8	3.5
Kohlrabi	(Bulbous) Stalk	90.3	2.0	6.6
Broccoli	Blossom buds and stalks	89.1	3.6	5.9
Cauliflower	Flower	91.0	2.7	5.2
Tomato	Fruit	93.5	1.0	4.3
Peas	Immature seed	78.0	6.3	14.6
Green Beans	Immature seed with pods	90.1	1.9	7.1
Lettuce	Leaves	95.5	0.9	2.9
Bean Sprouts	Young sprouts	88.8	3.8	6.6
Mushrooms	Whole (fruiting body)	90.1	3.1	3.8

[a] Source: C.F. Adams, Nutritive Value of American Foods, Agric. Handbook No 456, USDA, Washington, D.C.

span after the harvest season. This necessitates extensive and often elaborate raw material *storage* systems to allow for an economically viable processing season. The inherently high content of water in fruits and vegetables (Table 4-1) plays an important role in the economy of their storage for processing and merchandising. Any moisture loss during improper storage or in marketing may result in deterioration of quality because of wilting or drying out, and also in severe economic loss since most fruits and vegetables are sold by weight. The high water content of the fresh materials usually means high relative humidity of the storage atmosphere and this enhances the danger of *microbial spoilage* through action of bacteria, yeast, and molds. The commonly observed wilting tendency of some vegetables is related to their very high moisture content, sometimes in excess of 95 percent of the total weight (lettuce, cucumbers, green peppers). The capability of solid plant tissues to hold this much water is of great scientific interest. For the consumer it could mean "the most expensive water one can buy."

Since fresh fruits and vegetables are parts of *living plant* organisms, they are subject to various *biochemical processes* occurring during the plant life cycle and continuing after harvest. These biochemical changes are based largely on actions of enzymes, biochemical catalysts abundantly present in all fruits and vegetables. Many fruit and vegetable processing operations are designed to minimize the enzymatic deterioration of quality in storage, during processing, or in the final product.

4.2 PHOTOSYNTHESIS AND RESPIRATION

In essence, agricultural production of fruits and vegetables is based on a complex biochemical reaction known as *photosynthesis*. In gross oversimplification, photosynthesis can be visualized as the production of complex organic matter from water and carbon dioxide, aided by a supply of energy (light, heat) and catalyzed by chlorophyll and various plant enzymes:

$$6CO_2 + 6H_2O \xrightarrow[\text{chlorophyll enzymes}]{\text{heat light}} C_6H_{12}O_6 + 6O_2 \quad \text{(Equation 4-1)}$$

The resulting carbohydrates, such as the glucose shown in this simple example, may be further utilized in other biochemical reactions to produce proteins, lipids, vitamins and other chemical constituents of the plant tissue. The excess plant constituents will be stored in various parts of the plant; the predominant compound synthesized and the storage mechanism followed will vary substantially among individual plants. Thus, potatoes produce starch which is stored in tubers; peas are rich in protein, apples in sugar, avocados in fat, and kiwi fruit in vitamin C.

After the plant or its desired part is harvested, a reverse reaction called *respiration* will become predominant, providing the energy needed to sustain the ongoing life of the plant material. Again in principle only, this can be illustrated by Equation 4-2:

$$C_6H_{12}O_6 + 6O_2 \rightarrow 6CO_2 + 6H_2O + \text{energy (heat)} \quad \text{(Equation 4-2)}$$

The respiration reaction is vitally important for fruit and vegetable storage as will be discussed later.

4.3 INDUSTRIAL HARVESTING AND RIPENING

It is often said that food processing takes over where agriculture leaves off. However, modern industrial processing of vegetables and especially fruits is now inseparable from such a typical agricultural operation as harvesting. The rapidly escalating costs and scarcity of manual labor, on which harvesting used to be completely dependent, have catalyzed development of mechanical harvesting procedures for fresh produce. Today many delicate fruits and vegetables (strawberries, tomatoes, peaches, cherries, grapes) are being harvested by tree-shaking machines or by self–propelled mechanical harvesters (Figure 4–1) reminiscent of a similar recent innovation—the wheat harvesting combine which revolutionized the grain farming of the prairies. The mechanical harvesting of fragile, succulent fruits from perennial trees or plants, however, is a much more difficult task than harvesting grain. New biotechnological approaches had to be developed in collaboration with plant biochemists, breeders and engineers. The principal difficulty to be overcome was the impact damage of the soft produce. As softness is usually related to ripeness and full maturity, mechanical harvesting of less mature fruits became common, followed by final maturation in ripening chambers. It has been discovered that *ethylene* gas plays a role in the natural ripening process; the same gas used at appropriate concentrations (1–5 ppm) in the controlled conditions of ripening chambers will ripen green tomatoes, green bananas or mature but firm peaches in less than 10 days. New varieties of many fruits and

FIGURE 4–1. Mechanical harvesting of peaches. (Photo courtesy of Agric. Engineering Dept., Clemson University, South Carolina.)

vegetables are being developed by plant breeders to better withstand the rigors of mechanical harvesting at a more mature state. Various aspects of the mechanical harvesting technology for fresh fruits and vegetables continue to be developed as a response to the ever-rising costs of agricultural production.

4.4 FRUIT AND VEGETABLE STORAGE

Industrial processing of fruits and vegetables has developed much the same way as other food processing operations discussed in this text: from home processing and preservation of the excess products available after harvest to very sophisticated, computer-assisted operations. Some of the most important industrial fruit and vegetable products like jams, jellies, fruit juices, canned and frozen fruits and vegetables, potato chips, or tomato ketchup evolved from the same common need to utilize and preserve the macronutrients and at least some of the micronutrients produced by our efficient agriculture in a relatively short growing season.

Storage of raw material is one of the integral aspects of all industrial food processing. In the fruits and vegetables industry, proper storage control is one of the key prerequisites for successful marketing or further processing of fresh produce. After harvest, the metabolic activity of the living plant will continue in the harvested tissues causing rapid deterioration of quality and loss of valuable nutrients if storage conditions are not adequate. Especially important is the control of the *respiration* reaction (see Section 4.2). As shown in Equation 4-2, respiration of fruits and vegetables in enclosed storage areas will result in generation of CO_2, H_2O and heat. Proper ventilation to supply the required O_2 and temperature control to remove the heat generated by the respiration may be necessary to avoid quality changes of raw materials in storage. Inadequate oxygen supply might result in anaerobic respiration, in which the energy needed by the living tissues is obtained by breakdown of the constituent sugars to alcohol without O_2:

$$C_6H_{12}O_6 \rightarrow 2C_2H_5OH + 2CO_2 + \text{energy} \qquad \text{(Equation 4-3)}$$

While the anaerobic respiration produces the same end-products as obtained by yeast fermentation of many fruits (see Chapter 8), the two phenomena are fundamentally different and should not be confused. Anaerobic respiration is not a frequent phenomenon in industrial conditions although attempts are being made to utilize it in production of alcoholic beverages without the microbial fermentaton step.

An efficient control of the respiration reaction may be achieved by manipulations of the gaseous environment of the stored materials. In particular, modifying the amounts of O_2 and CO_2 in the storage atmosphere will affect the respiration rate; increased CO_2 content may reduce the respiration activity by two-thirds or more for some products. The *controlled atmosphere* (CA) storage concept has been developed successfully for prolonged storage of several commodities, particularly fresh apples and pears, assuring a practically continuous supply of these fruits year-round. The equipment needed for the proper control of temperature, relative humidity and the required gas composition in the atmosphere is rather sophisticated; the rooms must be airtight and proper air movement must be assured. The response of individual fruits and vegetables to the CA conditions is highly variable. Specific conditions have been established for the most common materials stored under CA conditions as shown in Table 4–2.

In home storage of lettuce, carrots, and certain fruits (particularly apples), a makeshift CA storage can be created by enclosing the fresh produce in polyethylene bags or other containers with low gas permeability (Chapter 11). The respiration reaction of the produce inside the bag will gradually reduce the O_2 levels while increasing the CO_2 content. The dangers of anaerobic respiration or microbial spoilage (enhanced by the excess H_2O produced in the respiration) must be controlled by keeping the container in a cool environment and checking periodically for signs of visible spoilage. When controlled properly, this simple arrangement may be useful in prolonging the storage life of fresh fruits and vegetables from the backyard garden.

Sometimes a carefully controlled storage treatment may be an integral part of the subsequent technological process. In potato processing, freshly dug potatoes must be provided with large amounts of O_2 for about five to seven days. In this early storage period the tubers that were damaged during the harvesting process will develop a secondary skin protecting them against rotting or microbial spoilage (a phenomenon called *suberisation*).

Table 4–2 Controlled Atmosphere Storage Conditions for Selected Fruits

	CO_2 (%)	O_2 (%)	TEMPERATURE (°C)
Apples, Jonathan	3.0–5.0	2.5	0.0
Apples, Golden Delicious	1.5–3.0	2.5	−1.1
Apples, McIntosh	5.0	2.5	2.2–3.3
Pears, Bartlett or Anjou	1.5–2.0	2–3	−1.1

Source: "Commercial Storage of Fruits and Vegetables", Agriculture Canada 1974, publication 1532. Courtesy Dr. W. Andrew, Dept. Plant Science, Univ. of Alberta.

For prolonged winter storage, the potatoes are kept in the darkness at low temperatures (4–5°C) and high relative humidity (80–90%). However, this favors a slow breakdown of the potato starch to sugar. For the manufacture of potato chips, French fries and similar products, high sugar content would result in excessive browning upon frying, which can be avoided by a *reconditioning* treatment. Reconditioning consists of storing the potatoes just before processing at 15–25°C for one to three weeks; this will accelerate various metabolic reactions resulting in using up the sugars, and/or their conversion back to starch.

Sophisticated storage systems may be used in industrial plants to prolong their processing season. A process has been developed to allow for year-round processing of tomatoes into ketchup and other products. The tomatoes are preprocessed by crushing the fresh tomatoes, heat-sterilizing the liquid pulp by a UHT process (Chapter 10) and storing it in large silo tanks under sterile conditions. A rail-car transportation system for tomato and other juices or purées for further processing is also available, based on similar principles. Storage systems for delayed fruit juice processing may utilize the much simpler pasteurization treatment (Chapter 6) since the high acidity of many fruits inhibits microbial spoilage. As a general rule, low temperature, high humidity and some protection from light are the minimum requirements for adequate preprocessing storage of fresh fruits and vegetables.

4.5 PREPARATIVE PROCESSES

Because of the origin and the nature of most fruit and vegetable raw materials, preprocessing operations such as washing, stone removal, sorting, peeling, and often mashing, slicing, cutting, chopping, pitting, and coring are common in most fruit and vegetable processing lines.

These preparative processes are predominantly of a mechanical nature, using various screens, rotating drums or knives, grinders and other fairly simple machinery. Inclined gravity feed conveyors, a characteristic sight in most industrial plants processing fruits or vegetables (Figure 4–2), connect these operations into continuous processing lines. Large amounts of water are used in the preparative processes for both cleaning and transportation purposes. The peeling of carrots, potatoes, beets, tomatoes, and the like can be accomplished by *mechanical abrasion* in rotating drums equipped with rough surfaces, quick exposure to *steam* followed by removal of the loose skins with water jets or other mechanical procedures, exposure to *hot lye* (10–20% NaOH at 80–90°C) with subsequent washing in a water tumbler, or *burning off* by flame (onions). The selection of the most appropriate system for a given pro-

FIGURE 4-2. An industrial processing line for French-fried potatoes.

cessing line will depend on the type of material to be processed, plant capacity, final products, availability and quality of water, waste disposal regulations in the surrounding community, and some other factors as discussed in Chapter 3.

4.6 BLANCHING

One of the most important characteristics of fruits and vegetables is their high enzymatic activity caused by the multitude of enzymes present. Blanching, or inactivation of enzymes by controlled heating, is a very important operation used in most vegetable and some fruit processing. The primary purpose of blanching is prevention of deteriorative enzymatic reactions resulting in browning of cut apples or potatoes, bitterness development in stored orange juice concentrate, development of off-flavors in frozen peas due to lipid oxidation, and many similar defects. Additional reasons for blanching include partial cooking, tissue softening, cleaning of cut surfaces, and other effects as summarized in Table 4-3. Selection of the appropriate blanching treat-

Table 4-3 Effects of Blanching in Processing of Vegetables

EFFECT	EXAMPLE OF USE
Deactivation of enzymes	All vegetable processing
Softening of tissues, partial cooking	All vegetable processing
Washing of cut surfaces	Potatoes for frying
Leaching of surface sugars	Potatoes for frying
Removal of air from tissues	Increased efficiency of canning or freezing (apples, green peas)
Pre-shrinking of material	Canning of mushrooms (avoid "half-empty" cans)

ment may vary for various materials. The main factors affecting blanching are composition and textural characteristics of the fresh material, uniformity and particle size of the pieces to be blanched, the final product being manufactured (frozen French fries, canned vegetables, etc.), blanching medium to be used (most often hot water or steam), and the machinery available.

These factors influence the choice of the *time* and *temperature* of blanching—two most important determinants of a successful blanching process. As a very rough guide, two to five minutes at 90–95°C may be adequate for home blanching of garden vegetables such as peas, cut beans, carrot cubes, etc. In industrial processes, specific blanching conditions are established for each product.

The simplest blanching technique useful for home or industrial processing is immersion in hot water for the required time. The higher the water temperature, the shorter the necessary time. Two or three blanchers in a series with carefully controlled temperatures may be used in industrial processing in order to save energy and avoid excessive loss of water-soluble nutrients by leaching. While steam blanching will minimize the nutrient loss, steam is a more difficult medium to work with than water and appropriate safety precautions must be utilized. One of the latest innovations in blanching technology is the Canadian–developed "K-2" blancher (Figure 4–3) which combines all the major advantages of steam blanching—the energy economy, the low nutrient losses, and negligible effluent discharge—in a safe and efficient design.

Heat resistance of individual enzymes that have to be inactivated by blanching varies greatly with the type of enzyme and also from vegetable to vegetable. Some of the common enzymes important in fruit and vegetable processing are listed in Table 4–4. An effective blanching process must deactivate all enzymes of significance for specific quality changes. Therefore, a test for the presence or absence of the most heat-

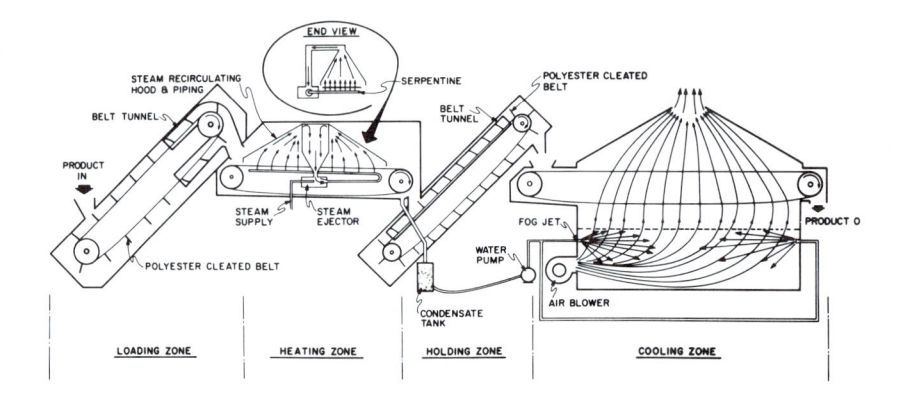

FIGURE 4-3. A schematic illustration of the "K-2" blancher principle and a photograph of the industrial machine. (Courtesy of Dr. G. Timbers, Engineering and Statistical Research Institute, Ottawa, Canada.)

resistant enzyme in blanched materials can be used to determine the adequacy of the blanching treatment. The enzyme *peroxidase* is usually employed for this purpose as an indicator of sufficient blanching. The effectiveness of a given blanching treatment in each processing situation may be controlled using inactivation of peroxidase, which is present in most plant materials. This is one of the most heat-resistant enzymes of plant origin, and is relatively easy to test for. Other enzymes may be preferred by industrial processors for various reasons. There is no regulatory requirement specifying the control procedure, unlike the

Table 4–4 Examples of Enzymatic Activity in Some Plant Materials

ENZYME	REACTION	EFFECT
Lipoxygenase (lipase)	Fat hydrolysis	Development of rancidity in raw milk, soybeans, frozen peas.
Polyphenol oxidase	Oxidation of phenolic compounds (catechol)	Enzymatic browning in apples, potatoes, pears, bananas.
Pectinase	Breakdown of pectin	Inability to produce jam or jelly.
Maltase	Breakdown of starch	Production of maltose in barley for brewing.
Chlorophyllase	Breakdown of chlorophyll	Solubilization of green pigment from spinach and other green vegetables; discoloration.
Peroxidase	Cleavage of oxygen from hydrogen peroxide	Used as a test for blanching adequacy.

enzymatic deactivation test required for determination of pasteurization adequacy in dairy processing (Chapter 6).

As an alternative to blanching, treatment with certain approved food additives is sometimes used for chemical blocking of enzymatic reactions in situations where heat treatment would be damaging. Browning of raw peeled potatoes destined for delayed use in restaurants, cafeterias and similar mass feeding establishments can be prevented by a dip in a solution of sodium bisulphite. Sulphur dioxide produced by burning sulphur may be used to prevent excessive browning during sun drying of apricots, peaches and other fruits. The use of approved chemical preservatives is permitted in frozen mushrooms as an alternative to blanching which would damage their delicate texture thus making freezing of mushrooms very difficult. Normally, vegetables must be blanched prior to freezing to control microbial contamination and enzymatic activity which is still possible in the frozen state and could result in quality deterioration (although not necessarily a health hazard) in the final product.

4.7 FRUIT AND VEGETABLE PRODUCTS

Further processing of fruits and vegetables is based primarily on the long-term preservation principles as discussed in Chapter 10. Freezing, drying, canning, pickling, production of concentrated products with or without added sugar or their combinations are all used to extend the shelf-life of these highly perishable commodities. In some cases, the processes are designed to preserve the fruits and vegetables in a form that is not substantially different from the fresh material (frozen or canned vegetables, dried fruits, pickled cucumbers, strawberries frozen in sugar syrup). Causes of some unavoidable quality changes, including texture softening effects of freezing or canning, toughening of texture in drying, lipid oxidation in dried products, or minor nutritional losses are explained in Chapter 10.

Other techniques have been developed to transform fruits and vegetables into a variety of nutritious, delicious, and convenient products. Frozen, ready-to-heat *French-fried potatoes*, or the much maligned *potato chips* are very important commodities both economically and nutritionally. Frozen French fries are usually *par-fried* (partially processed) so that in reheating the product at the point of final consumption, the necessary heat will not result in overcooking. The institutional use of frozen par-fried potatoes, the mainstay of many fast-food chain operations, is a significant example of market opportunities available to food processors in modern food distribution systems. Since color is an important sensory attribute of French fries and potato chips used by the consumer to perceive their overall quality, various technological means are included in the processing sequence to obtain the most desirable appearance. In addition to reconditioning (see Section 4.4) and blanching (Section 4.6, Table 4–3), the uniformity of the final color may be controlled by dipping the blanched potato slices in solutions containing glucose or starch, and by careful time and temperature control of the frying time. Nutritionally, fried potatoes are a valuable, seldom recognized source of vitamin C (Table 4–8).

Most fruits contain high amounts of water and are suitable for manufacturing of *fruit juices*, another important product of the fruit and vegetable industry. The relatively simple processing technology is based on mechanical pressing. Because of their high organic acid content, fruit juices have desirable thirst-quenching properties (Chapter 8). From a nutritional standpoint, ascorbic acid (vitamin C) is the most important acid present. The high amount of simple sugars in grape and many other fruit juices makes them ideal for alcoholic fermentation and use by the wine industry (Chapter 8). The shelf life of fruit juice products can be increased by relatively mild heat treatment; the high acidity itself prevents most spoilage microorganisms from prolific growth. Long-term preservation and substantial improvement in the economy of transportation and marketing can be achieved by partial removal of the constituent water by vacuum (and thus low temperature) evaporation, addition of sugar, and freezing. Frozen concentrated orange juice has become the single most important processed fruit product in terms of quantities of fresh fruit consumed.

While freezing is the primary preservation principle of sweetened fruit juice concentrates, the use of sugar alone can result in prolonged shelf stability of fruit products. In contrast to the historically important but nowadays market-restricted sugared (candied) whole fruits and fruit pieces, the production of jams, jellies and preserves is by far the most important representative of the sugar preservation technology today. The principle of jam and jelly production is simple, but the technological complexities of the process and the resulting multitude of legally defined products deserve further explanation.

4.8 JAMS AND JELLIES

The preservation principle used in these products is to reduce water activity by adding rather substantial amounts of sugar, thereby decreasing the potential for microbiological spoilage (Section 2.3). Jams and jellies are well defined in food laws of various countries. Common specifications include the minimum proportions of fruit and sugar allowed (typically 45 parts of fruit and 55 parts of sugar) and the required amounts of total water soluble solids (65–66% for jams, 60–62% for jellies) in the finished product. Some of the specification limits for the various products available on supermarket shelves are illustrated in Table 4–5.

Table 4–5 Examples of Compositional Differences of Jam and Jelly Products[1]

PRODUCT	COMPOSITION[2] % FRUIT	% SOLUBLE SOLIDS
Jam	45 (52[a])	66
Jam with pectin	27 (32[a])	66
Jam with apple or rhubarb	12.5 + 20[b]	66
Marmalade[c]	unspecified	65
Marmalade with pectin	27	65
Preserve	45[d]	60
Jelly	unspecified	62
Jelly with pectin	32	62

[1] Source: Canadian Food and Drug Act and Regulations, Ottawa.
[2] *Notes:*
  [a] For strawberries.
  [b] Apple or rhubarb pulp in addition to the fruit.
  [c] Denotes products made from citrus fruits only.
  [d] For every 55 parts of sugar before boiling.

Several distinctions among individual products are related to *pectin*, an important ingredient in jams and jellies. Pectin is a technical term applied to complex colloidal carbohydrate-like substances which occur in some fruits and which can form acidic gels with sugar and water (chemically, pectin is based on a naturally occurring polygalacturonic acid). The optimum pH limits for the gel-forming activity of pectin are 2.8–3.2; no gel will form above about pH 3.5, and below pH 2.5 the gel is too hard. Many fruits used for jam and jelly making contain enough organic acids to provide the optimal environment. For those that do not (e.g. peaches, cherries, pears), citric acid may be used to adjust the pH as required.

Some fruits, primarily apples and rhubarb, but also citrus fruits, cranberries, and sweet cherries, are rich enough in pectin to allow manufacture of jams and jellies of good consistency without further addition. Other fruits (strawberries, apricots) are poor sources of pectin and a commercial pectin preparation must be added. Inclusion of the word "pectin" in the descriptive name of any commercial product may signify that a lower proportion of fruit has been used and thus commercial pectin had to be added for an acceptable consistency (Table 4–5). Alternatively, jams can be made from mixtures of a given fruit with apples or rhubarb as one of the inexpensive ingredients to provide sufficient pectin. Addition of commercial pectin or acid in the highest quality jams and jellies is permitted only in reasonable amounts to compensate for any deficiency of the natural pectin content or acidity of the fruit.

The technology of industrial or home preparation of jams and jellies is relatively simple. The fruits are boiled first to soften the texture, to extract the flavor components, and to enhance the gel-forming properties of the naturally present pectin. For jelly making, the fruit is pressed to obtain a clear juice required as the fruit ingredient in jelly; whole crushed fruit is used in jams. Enough sugar is then added to satisfy the legal and technological requirements for sufficient solids content. Pectin and acid are also added at this point if required. The mixture is boiled until the desired solids content is reached. This can be accurately measured using a *refractometer*, a simple instrument (Figure 4–4) based on the known phenomenon of light being refracted as it passes through water (the familiar "bent-spoon-in-a-glass-of-water" effect). If dissolved substances are present in the water, the angle of "bending" (refraction) will change in exact proportion to the amount and type of solute present. As the principal solute in jams or jellies is sugar, the measurements of refractive index give a relatively accurate estimate of the total soluble solids content. Table 4–6 gives selected refractive index data for sugar solutions in the concentration range of

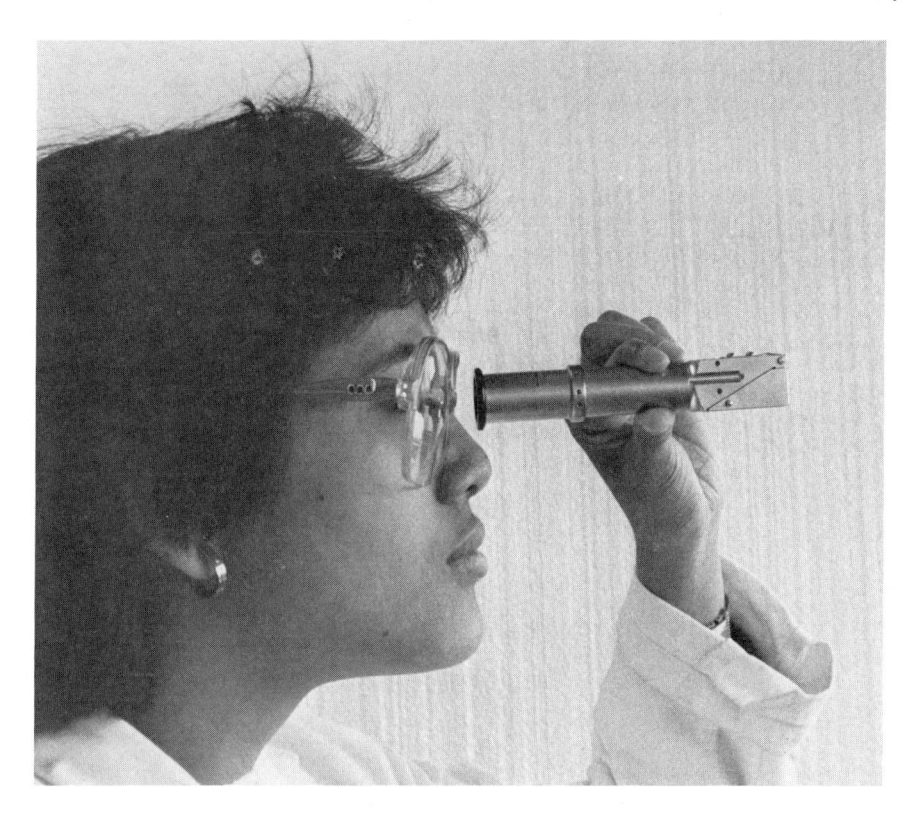

FIGURE 4-4. A portable refractometer used for process control in jam and jelly manufacture.

jams and jellies. If a refractometer is not available, the approximate soluble solids content can be estimated by repeated measurements of the boiling point of the mixture being processed. At the final stage of processing the boiling point should be about 4–6°C higher than that of pure water at the given location. The reason for the phenomenon of boiling point elevation is the same as that of the freezing point depression described in Chapter 10. Filling the hot finished product into clean containers with air-tight lids should result in proper gelatinization upon the slow cooling, and a suitable shelf life in conventional storage without refrigeration. For optimum consistency the final product should have a pH of about 3.2, and should contain about 65–66% total solids (°Brix in industrial terminology). Jellies of less than 60 °Brix will be weak or the gel will not form; products having more than 70% solids are likely to develop grainy texture due to the development of sugar crystals in these highly saturated solutions.

86 Processing of Fruits and Vegetables

Table 4-6 Refractive Index of Solutions of Sucrose in Water (20°C)

PERCENT SUCROSE	REFRACTIVE INDEX
0.0	1.3330
5.0	1.3403
10.0	1.3479
20.0	1.3639
30.0	1.3811
40.0	1.3997
50.0	1.4200
55.0	1.4307
57.5	1.4362
59.0	1.4396
60.0	1.4418
61.0	1.4441
62.0	1.4464
63.0	1.4486
64.0	1.4509
64.5	1.4521
65.0	1.4532
65.5	1.4544
66.0	1.4558
67.5	1.4593
70.0	1.4651

4.9 MANUFACTURE AND CRYSTALLIZATION OF SUGAR

Although processing sugar cane or sugar beet into white granulated sugar is not normally considered part of the fruits and vegetables industry, there are many similarities including production of the raw materials and their compositional characteristics. This relatively complex and specialized technology is based on the same principle as the graining in jams—granulated sugar is produced by crystallization from a highly concentrated sugar beet or sugar cane juice.

The typical sugar content of fresh juice extracted by pressing, cutting and leaching raw materials with water is about 10 percent. Thus subsequent evaporation of substantial amounts of water from this solution is necessary to exceed the sugar solubility limit and to create a *supersaturated solution*, which is a prerequisite for any crystallization to occur. The solubility of sucrose in water is highly dependent on temperature. The data for a pure sucrose-in-water solution are shown in Figure 4-5. Similar relationships are typical for solutions of other

4.9 Manufacture and Crystallization of Sugar 87

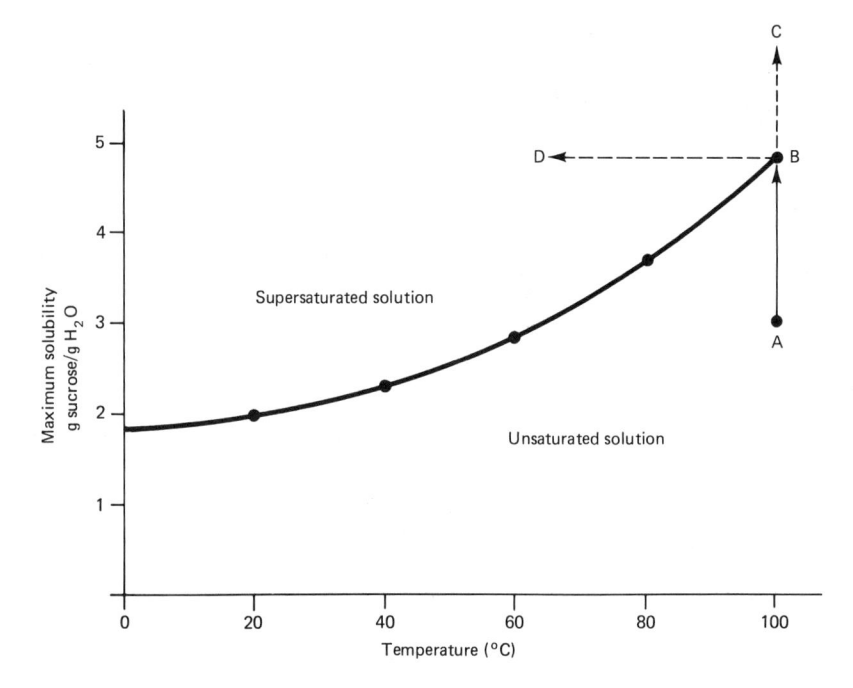

FIGURE 4–5. Solubility curve for sucrose in water and illustration of the formation of supersaturated solution. (Point A—initial conditions; Point B—saturated solution formed by boiling at 100°C; Point C—supersaturated solution formed by continued evaporation of water from solution B; Point D—supersaturated solution formed by cooling solution B.)

sugars in water, even though the solubility values themselves may vary widely. Table 4–7 gives solubility values of sucrose in comparison with other sugars important to the food industry.

The details of sugar processing technology are beyond the scope of this text; however, the basic principle is simple. Sugar is extracted from raw material in the form of juice, and a supersaturated solution is formed by evaporation of sufficient water, sometimes in combination with gradual cooling of the concentrated solution (Figure 4–6). Growth of sugar crystals is induced in the supersaturated solution, usually by adding very finely ground sugar crystals ("seeds"). The seed crystals will grow in the supersaturated solution as long as the supersaturation is maintained by continuing evaporation, cooling, or both. After the crystals have grown to a suitably large size, they are separated from the uncrystallized solution (*mother liquor*), which also contains various other dissolved organic and inorganic components ("impurities") from the raw material. Some of the impurities may remain on the surface of the raw sugar crystals. If so, these may have to be redissolved in clean

Table 4-7 Maximum Solubility (Saturation Values) of Selected Sugars in Water

TEMPERATURE °C	MAXIMUM SOLUBILITY (g SUGAR/100 g H$_2$O)		
	SUCROSE[a]	GLUCOSE[b]	LACTOSE[c]
0	179.2	51.0	11.9
10	190.5	67.9	15.1
20	203.9	89.6	17.0
30	219.5	120.1	23.2
40	238.1	163.8	32.0
50	260.4	229.7	44.2
60	287.3	284.0	50.5
70	320.5	359.3	77.5
80	362.2	440.2	99.0

Sources:
[a] Honig, P. *Principles of Sugar Technology*, Vol. 1 (New York: Elsevier Publ. Co., 1953).
[b] Young, F.E. (1957). J. Phys. Chem. Vol 61, p 616.
[c] Jelen, P. 1973., *Ph.D. Thesis*, from date of Whittier (J. Dairy Sci. 27, 1944:505.)

water and recrystallized, or *refined*. Without the refining step the raw sugar may be brownish in color due to the occluded mother liquor usually referred to as *molasses*. Certain components of the molasses are useful as human nutrients; however, the very small amounts of molasses present in the unrefined brown sugar have no practical significance in human nutrition.

4.10 NUTRITIONAL IMPORTANCE OF FRUIT AND VEGETABLE PRODUCTS

Fresh fruits and vegetables are our primary source of dietary vitamin C. Contrary to common public belief, modern processing techniques used by the fruit and vegetable industry do not result in large-scale destruction of this or other important micronutrients contained in the raw materials. Because vitamin C is not highly heat stable, any processing involving heat (including home cooking) will result in some vitamin C loss. Vitamin C is also water soluble and further losses by leaching may occur in liquids discarded after blanching, soaking, cooking or thawing previously frozen products. Table 4-8 compares the vitamin C content in several processed fruit and vegetable products with their fresh counterparts. As can be seen, the vitamin C is not completely destroyed even by the most severe canning processes. In some pro-

Table 4-8 Content of Vitamin C in Selected Fresh and Processed Fruit and Vegetable Products[a]

PRODUCT	FRESH	CANNED	FROZEN
	\multicolumn{3}{c}{VITAMIN C CONTENT IN mg/100 g}
Green peas	27	9	19
Green peas, cooked	20	n.a.	13
Orange juice	50	40	47
Potato	20	n.a.	9
Potato chips	16	n.a.	n.a.
Green peppers	105	68[b]	n.a.
Tomatoes	21	17	n.a.
Tomato ketchup	n.a.	15	n.a.
Peaches	6	3	40[c]
Kiwi fruit	120	n.a.	n.a.[d]

[a] Data from "Nutritive Values of American Foods" by C.F. Adams, Agric. Handbook No 456, USDA, Washington, and other sources.
[b] Chili sauce
[c] Ascorbic acid (vitamin C) added as a preservative against browning.
[d] Not available

cessed products, such as frozen peaches, the ascorbic acid content may be higher than in the raw material due to its technological use as a preservative against browning.

Vitamin A and fiber are other nutritionally important components of our diets supplied abundantly by some fruits and vegetables. Leafy vegetables used fresh (salads, cole slaw) or processed (sauerkraut, cooked spinach) are especially suitable as natural dietary sources of fiber; these "side dishes" should constitute a regular component of our daily meals.

There are no significant nutritional complications resulting from consumption of fruits and vegetables. The high oxalic acid content of spinach, Swiss chard, beet leaves, or rhubarb is sometimes mentioned in conjunction with impairment of calcium absorption due to formation of an insoluble salt, calcium oxalate. This is not considered to be of major nutritional importance as long as a sufficient supply of Ca in the diet is assured from sources that are much more significant than the fruits or vegetables (primarily milk and dairy products—Chapter 6). The highly insoluble calcium oxalate may be one of the primary compounds of certain types of kidney stones; thus it may be actually advantageous to combine the intake of these leafy vegetables with milk to block the absorption of the soluble oxalic acid. A more undesirable effect of some fruit products (jams, compotes in heavy syrup) or vegetables processed with sauces may be derived from the extra calories contributed by the added sugar or fat. However, because most of these foods are eaten in relatively small amounts—often as desirable components of a more complex meal, this too may not be particularly worrisome. Overall, fruits, vegetables and industrial products made from them are among the most recommended components of our daily diets. They contribute few extra calories, some important micronutrients, and make consumption of macronutrients from other sources more enjoyable.

4.11 CONTROL QUESTIONS

1. Explain the biochemical process of respiration and the difference between aerobic and anaerobic respiration.
2. Mr. Tom A. Toe, a large ketchup manufacturer, wants to store fresh tomatoes for several months to prolong his processing season. He is considering several storage options for installation in his plant.
   a) Cold storage with adjustable temperature control but no humidity control.
   b) Dry storage at room temperature with relative humidity controlled at 15%.
   c) A sealed cold room kept at 5°C and 90% relative humidity with no ventilation.

   Which of the three options (if any) would you recommend as most suitable for the required purpose? Are there any other storage options that Mr. Toe should consider?
3. How much sugar is needed for making strawberry preserve conforming to legal specifications if we have 18,000 kg of fresh strawberries? What other ingredient(s) may be included and why?
4. In a sugar beet processing factory, sugar is being crystallized from the concentrated juice at 80°C. Using data for sucrose solubility in water, explain why a 75% (w/w) sucrose solution would not form any crystals. What has to be done for the crystallization process to begin?
5. What is enzymatic browning and how is it prevented?
6. Explain the principle used in determination of total solids in fruit products by refractometer. Would it be possible to use the refractometer for other foods, e.g. pickled cucumbers? What is 1 degree Brix?

7. Why is vitamin C content of commercially frozen peas lower than that of fresh raw peas from a garden? Would you expect vitamin C content of domestically frozen peas (using a household freezer) to be higher than that of the commercial product?

Suggested reference books—
Fruits and Vegetables

Bianchini, F., Corbetta, F., and Pistoia, M. 1973. *The Complete Book of Fruits and Vegetables.* New York: Crown Publ. Inc.

Braverman, J. B. S. 1963. *Introduction to the Biochemistry of Foods.* Amsterdam: Elsevier Co.

Duckworth, R. B. 1966. *Fruit and Vegetables.* Toronto, Ont.: Pergamon Press.

Lopez, A. 1981. *A Complete Course in Canning, I and II.* Baltimore: The Canning Trade.

Luh, B. S. and Woodroof, J. G. 1975. *Commercial Vegetable Processing.* Westport, Conn.: AVI Publ. Co.

Meyer, L. H. 1975. *Food Chemistry.* Westport, Conn.: AVI Publ. Co.

Nelson, P. E., and Tressler, D. K. 1983. *Fruit and Vegetable Juice Processing Technology.* Westport, Conn.: AVI Publ. Co.

Pantastico, E. B. 1975. *Postharvest Physiology, Handling and Utilization of Tropical and Subtropical Fruits and Vegetables.* Westport, Conn.: AVI Publ. Co.

Smith, O. 1977. *Potatoes—Production, Storing, Processing.* Westport, Conn.: AVI Publ. Co.

Wills, R. B. H., Lee, T. H., Graham, D., McGlasson, W. B., and Hall, E. G. 1981. *Postharvest — An Introduction to the Physiology and Handling of Fruit and Vegetables.* Westport, Conn.: AVI Publ. Co.

Woodroof, J. G., and Luh, B. S. 1975. *Commercial Fruit Processing.* Westport, Conn.: AVI Publ. Co.

5

Cereal Grains and Oilseeds

5.1 RAW MATERIALS

Agricultural products belonging to this commodity group include dry, mature seeds of several types of plants. There are two typical subgroups—certain cultivated *grasses* (wheat, barley, oats, rice, corn and other cereals), and *oil-bearing plants* (soybean, rapeseed, sunflower, cottonseed, peanuts, palm). Dry *legumes* (dry peas, lentils, various types of beans, etc.) may be sometimes included since their processing is often similar to grain-handling techniques (one Canadian example is the milling of peas). Conversely, immature legumes (green peas or beans) or grass seeds (corn) are more properly discussed under vegetables.

The common characteristic of these materials is their low moisture content (Table 1–3). As a result, storage requirements for grains and oilseeds are rather simple, including primarily protection from moisture, insects and rodents. This is one reason why wheat has become the world's principal reserve food commodity.

The main industrial use of cereal grains is for milling into flour and further processing into bread, pastries and other bakery products, pasta products, and many other foods made from flour. Cereals are also used as raw materials for a variety of relatively new breakfast foods, for production of alcoholic beverages (beer, whiskey, sake and other products discussed in Chapter 8) and for direct consumption (rice, corn).

The principal use of oilseeds is as a source of edible oils. Many of the seeds also contain substantial amounts of protein in addition to the oil, and the protein processing has become another important segment of the modern cereal and oilseed technology. The processing systems for the various industries are based on several different engineering and scientific principles such as mechanical grinding, liquid extraction, heat coagulation or physical and physicochemical texturization. In particular, important differences exist between the primary technologies for cereal grains and oilseeds processing, and these will be discussed separately for the two sub-groups using a few selected commodities as examples.

5.2 CEREAL GRAINS: KERNEL STRUCTURE AND MILLING

Milling of cereal grains to produce flour is one of the oldest food processing techniques developed by mankind. In the past, grinding stones driven by wind, water, or agricultural animals were used to crush the dry kernels. The modern flour mill is based on the same principle of mechanical disintegration of the grain between two heavy moving bodies. However, electricity-driven and computer-controlled stainless steel rolls replaced the nostalgic waterwheels of Germany or windmills of Holland (Figure 5-1), increasing the efficiency of the operation immeasurably.

Common wheat (*Triticum vulgare* or *T. aestivum*) is the most important cereal grown for milling into flour. Many varieties of wheat have been developed for different growing conditions (especially winter wheat, planted in the fall and harvested the following summer, vs. spring wheat planted and harvested in the same growing season). A similar differentiation can be made regarding various end uses: hard wheat (winter or spring) for bread-baking, soft wheat for cakes and cookies, and the very hard durum wheats for pasta products (macaroni, noodles, etc.). The USA, followed by Australia and Canada, is the major supplier of wheat in world trade (Table 5-1). Cereals (particularly wheat and rice), are probably the most important goods sold internationally. It is estimated that over one-half of all international trade is attributable to grain transactions. Processing of wheat, the most important Canadian and U.S. agricultural crop, will be used in this text as the primary example of cereal technology.

To understand the wheat milling process, one must be familiar with the composition and structure of the dehulled kernel (Figure 5-2). The outer protective layer, called *bran*, constitutes about 14 percent of the total kernel mass, and is composed primarily of fiber (cellulose). Since bran contributes to the indigestible bulk in food and has no positive effect on the dough-making properties of flours, it is usually removed in the milling process. Because a recent controversial medical hypoth-

FIGURE 5-1. An 18th-century windmill from Northern Europe.

Table 5-1 World Wheat Production and International Exports by the Main Wheat Producing Countries[a]

COUNTRY	TOTAL PRODUCTION[b] (MILLION TON)	INTERNATIONAL EXPORT[c] (MILLION TON)
USSR	90	0.5
USA	55	36.6
China	45	—
India	28	—
Canada	20	15.0
Australia	14	15.4

[a] Source: Canadian International Grains Institute, Winnipeg, Man.
[b] 1980
[c] 1979–1980

esis suggests a link between the lack of fiber in our diets and diseases such as cancer of the colon, diverticulosis, and cardiovascular disease, the total removal of bran from some flours is now being questioned. The bran and layers found immediately below it also contain some important vitamins of the B group and minerals (primarily iron) which are removed with the bran. Depending on the percentage of the bran re-

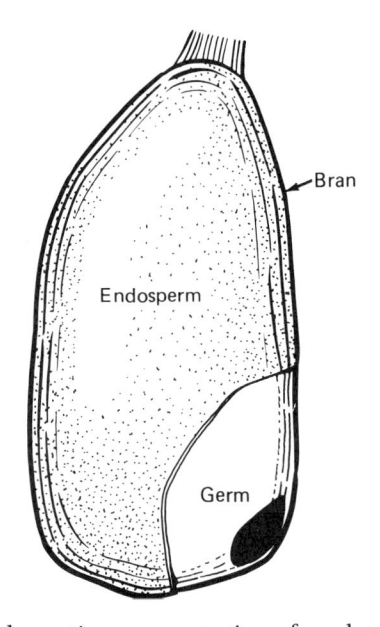

FIGURE 5-2. Schematic representation of a wheat kernel structure.

moved, flours of various *extractions* (denoting the percentage of total kernel mass remaining) are attained. The lower the extraction number, the more fiber is removed, and consequently the whiter the flour and the products made from it. Nowadays, even the whitest flours are supplemented with most of the vitamins and minerals removed in the milling process (thus, *enriched flours*). Extractions of common wheat flours are shown in Table 5-2.

The main portion of the kernel, known as *endosperm*, consists mainly of starch (approx. 70 percent) and protein (8-14 percent), and comprises about 85 percent of the kernel mass. The recovery and pulverization of the endosperm as flour or semolina (coarsely ground endosperm of the durum wheat) is the main objective of milling.

The third part of the kernel, *germ*, is the embryo from which a new plant develops upon germination. The germ constitutes only about 2 to 3 percent of the kernel, but as it is rich in fat (up to 10 percent of its total weight) it must be removed in the milling process, otherwise the flour would turn rancid during storage. The concomitant loss of protein (constituting up to 25 percent of the germ mass) is rather insignificant due to the very small total mass of the germ. The germ is also rich in thiamine, one of the vitamins added back to enriched flour. Even whole wheat and other high extraction flours prepared by modern milling methods have the germ removed. Government food laws usually specify that whole wheat flour "must contain not less than 95 percent of

Table 5-2 Main Types of Wheat Flours and other Wheat Products.

COMMERCIAL NAME	APPROXIMATE EXTRACTION (% OF TOTAL KERNEL)	APPROXIMATE CONTENT[a] OF PROTEIN %	CHO %	FIBER %
Whole flour	95	13.5	71	2.3
Straight flour	80	12	74	0.5
Patent flour (for bread-making)	72	11.8	75	0.3
All-purpose flour (family)	60-70	10.5	76	0.3
Cake and pastry flour	50-60	7.5	80	0.2
Gluten flour	—	42	44	0.3
Wheat bran	—	16	62	9
Wheat germ	—	27	47	2.5

[a] From various sources

the total weight of the wheat from which it is milled." (From the Canadian Food and Drug Act and Regulations, 1983.) The industrial milling process consists of a series of operations shown schematically in Figure 5–3. The initial *cleaning* steps are designed to remove impurities that may have been collected during harvesting, including seeds of various weeds, small stones, soil particles, and insect contaminants. In the *tempering* (conditioning) operation, the moisture content of the kernels is adjusted to cca 15 to 16 percent H_2O by adding a limited amount of water. This softens the bran without influencing the friability of the endosperm, allowing an easy separation of the two components after the seeds have been crushed by several pairs of *breaking rolls*. These stainless steel, corrugated rolls are interspersed with mechanical

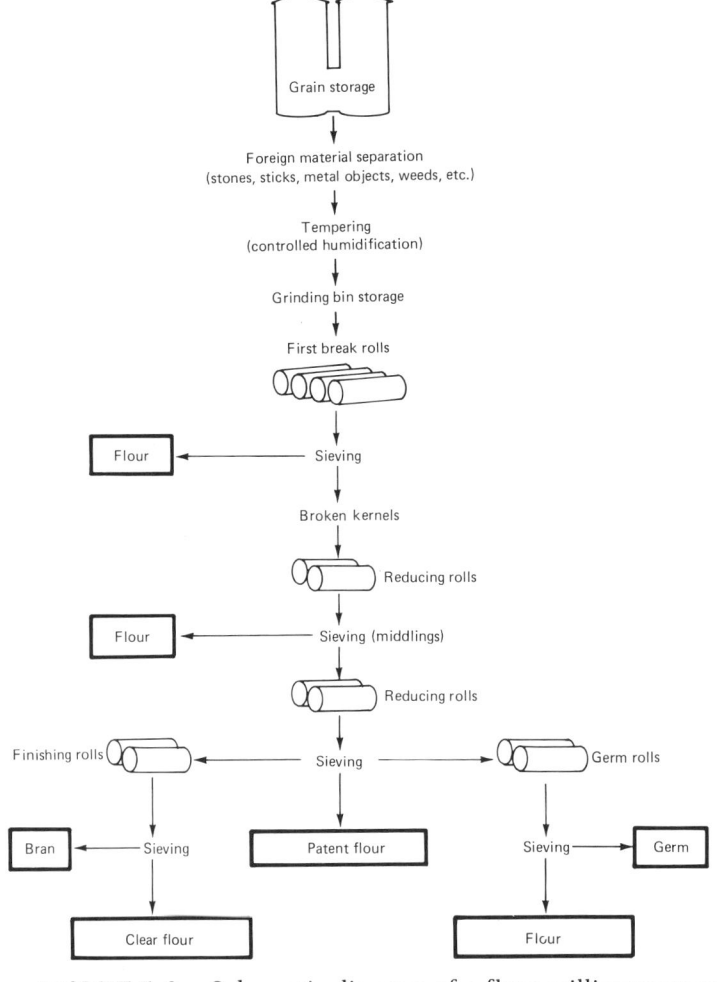

FIGURE 5–3. Schematic diagram of a flour-milling process.

sieves to divide the pulverized material into streams of various particle sizes. The coarse materials continue onto subsequent *milling rolls*, which remove more and more of the endosperm adhering to the bran. Further sieving removes the bran and germ from the middle streams (*middlings*) which are further pulverized and added to the main flour stream. When all-purpose or similar white flours of the typical 70 to 75 percent extraction are produced, the bran and germ are not used. For production of whole wheat flours, the separated bran is further pulverized and added back. As shown in Table 5-2, whole wheat flour differs from all-purpose flour by being slightly higher in protein (and also minerals), substantially higher in fat and fiber, and lower in starch contents. The difference in the fiber content may have some nutritional significance. Cereal foods such as bread and pasta are not our main sources of minerals or protein so the differences in these nutrients are unimportant. In addition, wheat protein is inferior nutritionally to other proteins since it is deficient in lysine and other essential amino acids. Technologically, however, protein is by far the most significant ingredient of wheat flour due to its unique dough-forming capability which is the basis for a successful baking process.

Bleaching and maturation of freshly milled flours are recent industrial innovations utilizing advances in knowledge of cereal chemistry. These steps accelerate chemical changes occurring naturally in prolonged storage which not only enhance the whiteness of the flour, but also improve its baking properties. Untreated, "green" flours are unsuitable for use in baking due to poor consistency of the formed dough resulting in inferior texture and generally low final quality of the baked products. Without bleaching, a storage period of up to six months may be necessary before the untreated flour can be used. Introduction of bleaching and maturation processes substantially reduced expensive storage, danger of insect infestation, development of rancidity, and other deteriorative changes typical of prolonged storage. Bleaching affects some vitamins present in flour, but the same is true for the natural maturation process. Various approved food additives, such as chlorine dioxide, potassium bromate, or ascorbic acid, are being used to accomplish the bleaching and maturation. The maximum permitted residual content of the bleaching agents in the flour (or in the bread produced from it) varies from 45 ppm - 2,000 ppm according to the individual chemical agent and the country concerned.

5.3 BAKING BREAD

Conversion of flour into bread is probably the most important food processing operation in the world. Bread has been identified as the source of more than 50 percent of the total food energy in more than 50 percent of all countries. The importance of bread as an indispensable source of food nutrients and energy has declined in Canada and the U.S.A., but baking is still one of the largest food industries in North America.

The principal steps in a typical bread baking operation include formation of dough, its fermentation, shaping and leavening ("proofing"), and finally conversion to bread by baking. The schematic sequence of events occurring during the bread-making process is shown in Figure 5–4.

The essential ingredients for bread dough are wheat flour, water,

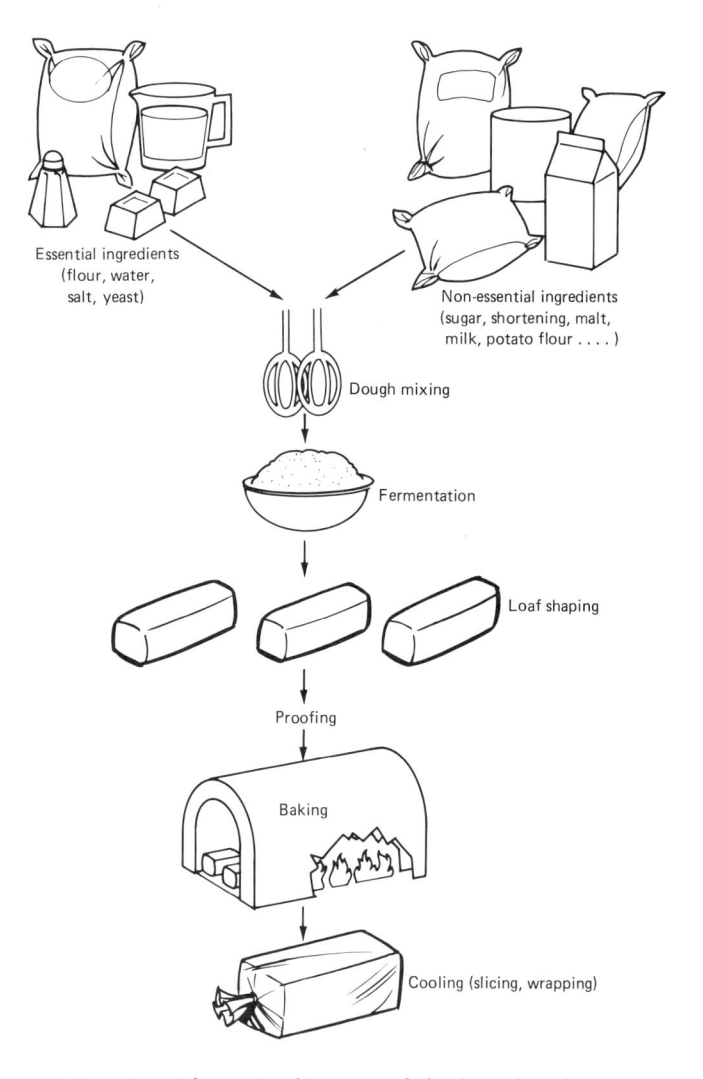

FIGURE 5–4. Schematic diagram of the bread-making process.

yeast and salt. Other additional ingredients such as rye flour, sugar, milk, potato flour, raisins, chopped onion, and other materials may be optional ingredients used in various specialty breads. Wheat flour—or more specifically, the wheat protein in the flour—is the most important component involved in the formation of dough. Wheat protein contains two principal chemically distinct components: glutenin and gliadin. When the wheat flour is mixed with water, the hydrated wheat protein complex (called *gluten*) gives the dough its elasticity, stickiness and pliability; this is essential in fermentation, loaf-making, and bread structure development upon baking. The superior bread-making property of hard wheat is its most distinctive property; no other cereal flour alone can be used to make a loaf of bread satisfying the traditional quality criteria of the western world.

The first step in the bread-making process is the formation of dough by mixing all the selected ingredients with water. After proper development by kneading, the dough is allowed to *ferment*. In this operation, the added yeast culture will convert simple sugars such as glucose, sucrose (present in the flour naturally) or maltose (produced by breakdown of the wheat starch by maltase or amylase enzymes from the flour), into CO_2 and ethyl alcohol. In the fermentation stage, the dough contains rather substantial amounts of alcohol, which is, however, all driven off by the heat applied in the baking step. Lactic acid bacteria (see Chapter 6) may also be used in addition to yeast for production of lactic acid from the added milk if *sour dough* bread is being made. The fermentation step is important for development of CO_2 gas which has a major effect on the final bread structure. Biochemical modifications of the gluten matrix during the fermentation process allow CO_2 retention during baking, and development of desirable flavor compounds results from the microbial metabolism also. The fermentation may be accomplished in the whole batch of dough prepared for bread (this approach is termed the *straight method*) or using only part of the total ingredients (*sponge method*) and later admixing the remainder of the required materials. There are many variants of the sponge method reflecting traditional differences in the art of bread-making in various countries. The sponge is usually more liquid than the final dough to optimize the fermentative activity. In general, the sponge methods allow for more flexibility of the baking schedule; unforeseen delays will not ruin the batch as in the case of the straight dough. Other claimed advantages include more uniform quality of the bread and some savings in the amount of yeasts needed as compared to the batch (straight) method. A major disadvantage is the increased need for labor and machinery.

After fermentation, the dough is formed into loaves. During this mechanical handling operation, much of the CO_2 is removed from the dough matrix. Final fermentation in the loaves (*proofing*) is necessary

102 Cereal Grains and Oilseeds

to insure proper volume and texture of the baked loaf. Required temperature (35–42°C) and relative humidity (85 to 90 percent) must be carefully controlled in the proofing operation to allow optimum yeast activity in this relatively short proofing time (45–60 min) before the loaves are moved into the baking oven. The proofing, as well as other bakery operations, may be carried out in batches or as a continuous process.

The main events occurring during baking are shown in Figure 5–5. The most important changes that take place during transformation of the dough to bread are the destruction of the microorganisms used in the fermentation; inactivation of the enzymes present in the flour; gelatinization of the wheat starch and heat coagulation of the gluten providing the bread structure; evaporation of the excess water from the dough; and darkening and toughening of the outer layer known as *crust*. Due to the continuous evaporation of water from the bread, the interior temperature does not exceed 100°C, hence the white color of the inner parts of bread (*crumb* in technical terminology). The final composition of breads is not specified by food laws and regulations;

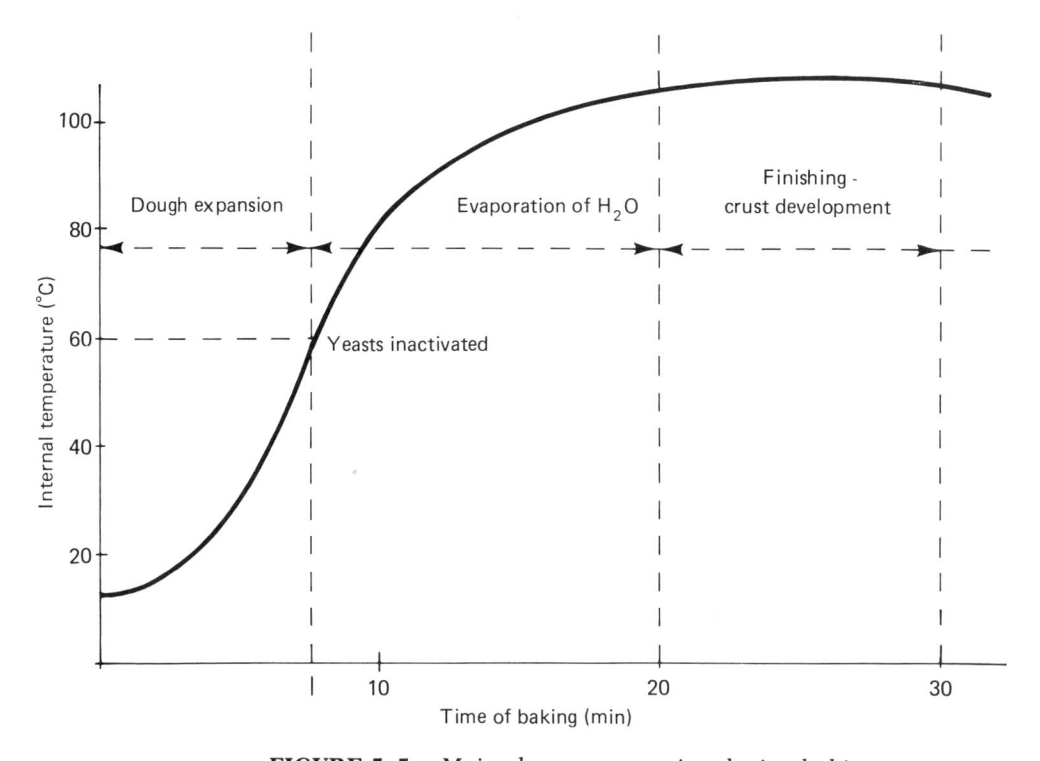

FIGURE 5–5. Main changes occurring during baking.

Table 5-3 Selected Compositional Data[a] for Several Types of Bread (Average Values)

BREAD TYPE	FLOURS USED IN ADDITION TO PATENT WHEAT FLOUR TYPE	% OF ALL FLOUR	FINISHED BREAD FOOD ENERGY (Kcal/100g)	H_2O (% w/w)	PROTEIN (% w/w)
White bread, enriched	—	—	270	35.6	8.7
Whole wheat	Whole wheat	50	245	36.4	10.5
Rye (sourdough or sweet)	Straight rye	20–30	245	35.5	9.1
Pumpernickel	Whole rye	12	245	34.0	9.1

[a] Source: Nutritive value of American Foods, by C.F. Adams, Agric. Handbook No 456, USDA, Washington, D.C.

orientation values for several common breads are in Table 5-3. On the average, typical white bread contains approximately 35 percent moisture.

The main quality factors considered important by bread makers are color, texture and uniformity of the crust and crumb, flavor, and the final loaf volume. In rye and other specialty breads, the acceptable loaf volume is insured by using at least 50 percent (typically 60 to 65 percent) of wheat flour. Rye flour protein does not provide for sufficiently strong gluten formation for acceptable final loaf volume. Dark heavy breads such as pumpernickel (whole rye) bread, some Scandinavian breads, and various oriental bakery products, are typical examples of the inadequate loaf volume development resulting from lack of wheat flour.

To improve the nutritional quality of breads (recall that wheat protein has poor nutritive value), various *composite flours* have been suggested. These may consist of wheat flour combined with materials rich in nutritionally adequate protein such as soy flour, potato flour, or fish protein concentrate. Breads made from composite flours usually have a better nutritional profile, however, the final product qualities mentioned above may be inferior. In particular, composite flours tend to depress the final loaf volume, produce a friable, incohesive crumb and generally give products of lower eating quality. A search for suitable composite flour ingredients continues to be one of the priorities of contemporary cereal science.

Packaging and storage of bread are simple but important components of the bread-baking process. Without proper protection, freshly baked bread will undergo a rather rapid change in texture known as *staling*. In addition to special ingredients (various dough conditioners and emulsifiers) used to maintain the desirable texture of fresh bread, packaging it in a moisture-impermeable plastic plays an effective role in ensuring quality preservation of the final product. Retaining moisture in the bread, by preventing evaporation from the package, is an important prerequisite for retarding staling. The moisture-impermeable packaging, however, results in development of high humidity inside the package which would enhance spoilage due to mold growth if a food-approved chemical preservative (propionic acid or its derivatives—see Chapter 9) were not used. Frozen storage, below $-2°C$, will stop the staling process and is an effective means for preserving fresh loaves of bread. After reheating in the oven, a "freshly baked" delicacy can be obtained. However, refrigerated storage of bread above its freezing point is ineffective and should be avoided as it will accelerate the staling reactions. In principle, staling is related to a complex chemical phenomenon of *starch retrogradation*, during which the hydrated, gelatinized starch granules (Figure 5-6) are converted to

FIGURE 5-6. An electron micrograph of starch granules in freshly baked bread. (Courtesy of Dr. D. Hadziyev, University of Alberta.)

a more permanent, hard structure. Without adequate precaution, staling of bread and other bakery products, together with their spoilage by molding, may result in significant wastage if these products are discarded.

5.4 PASTRY, COOKIES AND PASTA PRODUCTS

The preceding paragraphs may serve as a typical example of the milling—baking technological sequence employed generally for all bakery products.

Besides bread, wheat flour is required for the manufacture of "white pastries" such as buns, rolls, cakes, and many pastry specialities. The technological principles of baking as described above are usually applicable although different machinery is often needed. In European history the traditional use of highly refined flour of low extraction number ("fancy flour") in white pastry has been associated with foods used by the "higher classes." The highly refined but enriched flours available today are often criticized for lack of nutrients, however, Danish pastry, croissant rolls and similar culinary delights cannot be made without it.

Processes used for industrial (or home) production of cookies, crackers, doughnuts and other products of this type combine the principles of baking with elements of chocolate or confectionery technology. Some traditional bakery operations may be modified (oil-frying rather than baking of doughnuts) or omitted altogether (fermentation). In cookie technology the fermentation step may be replaced by *chemical leavening*, based on the use of sodium bicarbonate ("baking powder" or "baking soda", thus soda crackers). The action of water in the dough on the baking powder releases the needed CO_2 without the lengthy fermentation step, which is not required for the texture-formation in the cookie dough, and which would produce the typical yeasty flavor undesirable in the finished cookies.

Pasta Products

Noodles, macaroni, spaghetti and similar products are made from flour of the durum wheat varieties, which are hard and yellowish. Dough formation and processing for the pasta products follow a similar sequence of steps as in bread making, with the exception of leavening. The unfermented dough is formed into the desired shapes by various mechanical processes, and is then dried rather than baked. Boiling in water just before use will produce the starch gelatinization typical of cooked pasta products. Dumplings, the middle-European equivalent of the pasta, are boiled directly from the freshly made, often yeast-fermented dough without the drying step. Many variations due to the type of flour used, the shape of the final product, and creativity of the cook account for the almost endless selection of speciality dishes based on these products which are characteristic of some of the cuisines of Europe or the Orient.

5.5 MILLING AND PROCESSING OF OTHER CEREALS

Milling of rice and corn differs from the typical wheat milling process due to the different structure of grains and different end uses of the milled product. The main aspect of rice milling ("pearling") is the removal of only the husk, bran and germ, without the disintegration of the endosperm. Rice, which is predominantly consumed as the whole "milled" grain, is the single most important food commodity in the world.

In the milling of corn, the largest U.S. cereal crop, the primary objective is the recovery of cornstarch, although the germ is a very important by-product used for corn oil manufacture. Cornstarch has many industrial applications; among the most significant is the enzymatic conversion into corn syrup, a low-cost sweetener suitable for many processed foods.

Milling of rye, dried peas, or triticale (a man-made cereal developed by plant breeders as a cross between the industrially more desirable wheat and the agriculturally less demanding rye), is similar to wheat milling. The resulting flours are available for some specialty bread uses, of which rye bread (Section 5.3) is by far the most significant.

Special processes for modification of intact or partially milled cereal grains (flaking, puffing, application of dry heat as in pop-corn manufacture), are used for preparation of a variety of breakfast cereal products. Intact grains are also an important raw material for the manufacture of alcoholic beverages (Chapter 8).

Breakfast Cereals

The relatively recent development of ready-to-serve breakfast cereal products complements the traditional uses of wheat, oats, corn, rice and other grains in bakery and pasta foods. However, the idea of using cereals as a quick supply of energy in the morning is not new. The Scottish porridge of oats, the corn grits of New Orleans, or the traditional and universal toast, are all examples of breakfast use of cereals. Ready-to-serve breakfast cereal products are often fortified or combined with other sources of complementary nutrients. They are usually served with milk which provides the much needed calcium and phosphorus, and with fresh fruits as a source of vitamins. Obviously, the corn flakes, puffed rice, wheat dough rings and other recent innovations of modern food technology are just as valuable macronutrient sources (or "vehicles" for consumption of micronutrients from other sources) as the traditional croissant in the continental breakfast of Europe. The frequent criticism of presweetened cereals for their "empty calorie" content or even as "junk food" can be confronted with the often expressed argument that there are no junk foods—only junk eating habits, while the "empty calories" from the added sugar provide the quick energy that the human body needs in the morning. Furthermore, the unsweetened cereals are almost always sweetened before eating and it was shown that the amount of sugar consumed by children who add it to their cereals could be higher than when presweetened products are served.

The fundamental principle underlying the manufacture of breakfast cereal products is generally similar to baking, in that the starch of the grain materials is modified by heat to make it more digestible. According to the processing technique used, the amount and type of dry and moist heat may vary, and this may affect some of the heat-labile micronutrients. The actual technology used to produce the various flakes, puffs, rings and other bizarre shapes is quite sophisticated and often highly proprietary. It may include flaking of cooked grains by pairs of smooth rolls, breaking into bits and compressing together, or the new processing technique of *extrusion cooking*, where a dough is cooked as it is transported under high pressure through a double-jacketed heated "gun," much like a big meat grinder (Figure 5-7). The sudden release of the cooked dough to the atmosphere at the end of the cooking extruder then causes the superheated water to evaporate instantly, and this provides the puffed texture characteristic of many breakfast cereal products.

5.6 MALTING

Malted barley is a primary ingredient in beer-making. Malting may be considered a first step in the traditional brewing technology (Chapter 8) since many older breweries used to carry out their own malting processes. In North America almost all barley malt is produced by specialized *malt houses* and supplied to brewers around the continent. This, together with other technological reasons, may account for the relatively small differences between the various national brands of beer.

Malting is principally a three-stage process accomplished in large vessels and rooms. Its main purpose is to germinate the barley seeds so that the various dormant enzymes present in the seed will become ac-

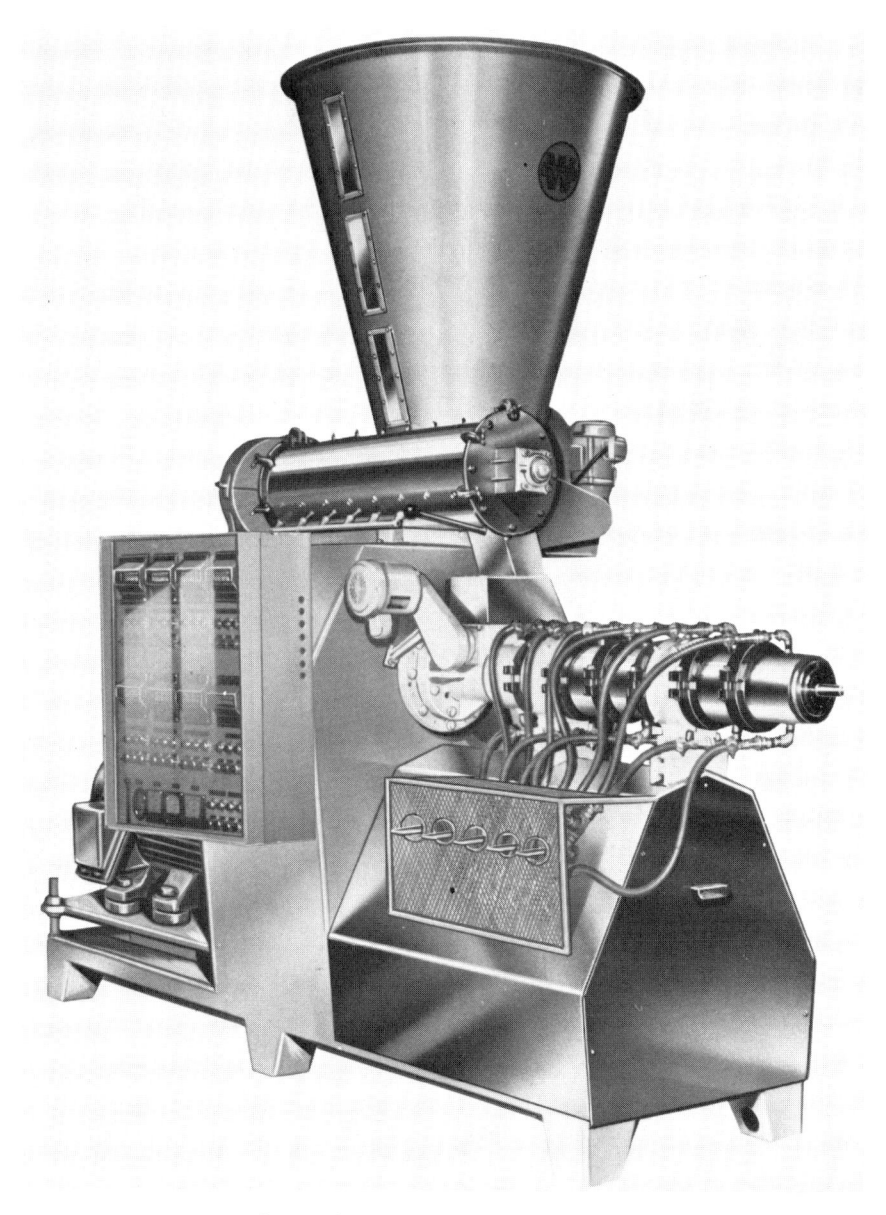

FIGURE 5-7. An industrial extruder-cooker for processing of cereals and other food products. (Courtesy of Wenger Manufacturing Co.)

tive. The most important enzymatic reaction required is the breakdown of the starch into the much smaller molecules of its structural component, *maltose*. This "cereal sugar" is later utilized by the yeasts in the brewing process to produce alcohol and carbon dioxide. For the germination to occur, the seeds must be soaked in water (*"steeping"*) until they contain approximately 40 to 44 percent moisture. This increases the seed volume by about 30 percent and results in a rapid respiration reaction (Chapter 4). Its main products, CO_2 and heat, must be continuously removed in the subsequent *germination* step and the O_2 supply, temperature, humidity and CO_2 concentration must be carefully controlled. The activated seeds are dried in the *kilning* step. This gentle drying by warm air blown through the beds of the germinated seeds stops the germination process without destroying the enzymes which will be revitalized once more in the brewing process. The dry malt is storage-stable and can be easily shipped across the country to various brewers. Further discussion regarding the use of cereals in alcoholic beverage production can be found in Chapter 8.

5.7 OILSEEDS: COMPOSITION AND EXTRACTION OF VEGETABLE OILS

The primary objective of oilseed processing is to recover the substantial amounts of edible oils contained in these rather unusual seeds. The main oil-bearing plants are listed in Table 5–4 with their approximate oil content and some yield data of interest. The typical oil content in the seeds may vary from low of about 20 percent for soybean to more than 50 percent in the sunflower. Despite its relatively low oil content (compensated for by high yields per acre of land), *soybean* is by far the world's most important oilseed today, and the U.S. (mainly Illinois, Indiana, Iowa and Ohio) is the principal producing country. The value of the soybean is not only in its high quality oil, but also as an important protein source for animal feed and, more recently, as a human food ingredient with special functional properties.

Due to more severe climatic conditions, the main oilseed crop in Canada (as well as in several European countries such as Poland, Sweden, Germany, Switzerland, Czechoslovakia) is *rapeseed*, belonging to the *Brassica* group of plants which also include cabbage, broccoli, mustard and other commonly grown agricultural crops. In spite of the long history of use of the traditional rapeseed oil in Europe, relatively recent medical findings indicate a possible connection of one of the principal fatty acids found in older rapeseed varieties, the erucic acid, to cardiovascular complications. The currently grown varieties of Canadian rapeseed are the result of successful breeding programs

Table 5-4 Selected Production[a] and Compositional Data for Major Oilseed Crops

	WORLD PRODUCTION (Mt)				
REGION	SOYBEAN	SUNFLOWER	PEANUT	RAPESEED (INCL. CANOLA)	COTTONSEED
North America	50.5	1.9	1.3	2.5	4.9
South America	19.5	1.7	0.9	0	1.8
Africa	0.3	0.5	4.8	0	2.1
Asia (without USSR)	11.9	1.3	11.9	4.2	8.8
Europe and USSR	1.1	7.5	0	3.6	4.6
Composition					
Oil (% w/w)	22	45	48	46	33
Protein (% w/w)	41	24	28	21	38
Average Yield (t/ha)	2.2	0.8	0.9	1.0	n.a.[b]

[a] Source: FAO Production Yearbook, Vol 34, FAO, Rome, 1980.
[b] not available

5.7 Oilseeds: Composition and Extraction of Vegetable Oils 111

which removed from the seeds not only the controversial erucic acid, but also the most undesirable components of protein meal (a group of chemicals known as glucosinolates). The generic term *canola* describes these new varieties sometimes referred to as "triple-O" (low in erucic acid, glucosinolates and the dark bitter pigment of the seed) which are grown almost universally in Canada today. The success of the canola breeding program is another example of the interdependence of modern food processing and the agricultural science.

The basic components of vegetable oils (as well as animal fats and other lipids) are various *fatty acids* which are bound in mixed proportions to single molecules of glycerol, a compound related to glucose (Figure 5-8). Numerous fatty acids can be part of any particular oil; varying contents of the individual fatty acids differentiate the individ-

Figure 5-8 Schematic representation of saturated (a), unsaturated (b) fatty acids and of a mixed triglyceride (c).

a. Palmitic Acid:

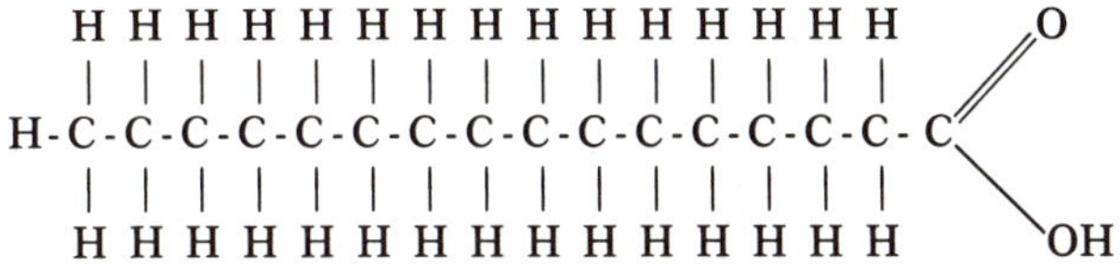

$$H-C-C-C-C-C-C-C-C-C-C-C-C-C-C-C-C(=O)OH$$ (with all H's on each C, 15 CH's + COOH)

b. Linoleic Acid:

$$H-C-C-C-C-C=C-C-C=C-C-C-C-C-C-C-C-C(=O)OH$$

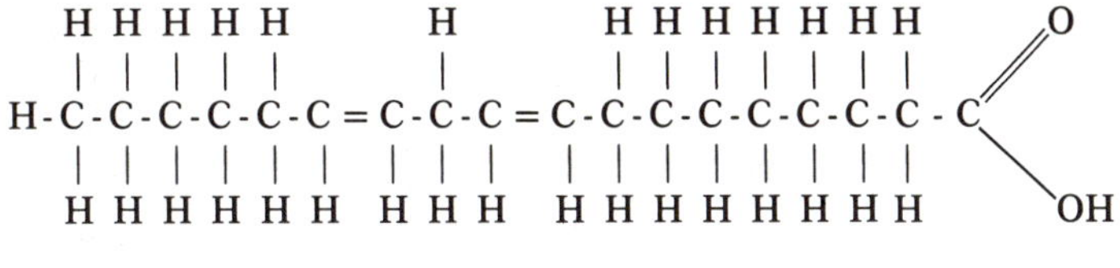

c. Mixed Triglyceride with three different fatty acids attached to one molecule of glycerol.

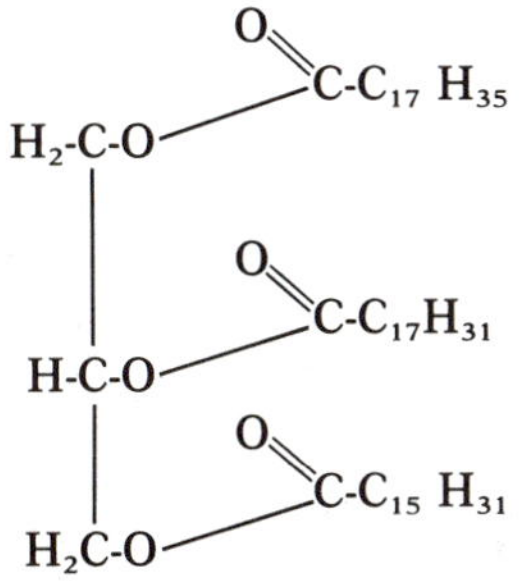

$$H_2C-O-C(=O)-C_{17}H_{35}$$
$$H-C-O-C(=O)-C_{17}H_{31}$$
$$H_2C-O-C(=O)-C_{15}H_{31}$$

Table 5-5 Content of Selected Fatty Acids in Some Sources of Edible Lipids[a]

LIPID SOURCE	SATURATED FATTY ACIDS	OLEIC ACID (18:1)	LINOLEIC ACID (18:2)
	---------------- % total fatty acids ----------------		
Coconut	85	6	2
Peanut	16	55	21
Olive	15	47	8
Soy	13	30	52
Canola[b]	8	56	24
Butterfat	63	30	3
Beef Tallow	55	26–50	2
Pork Fat	40	47	8
Fish Oil (Herring)	19	25[c]	

[a] Principal Source: Handbuch der Lebensmittelchemie, Vol 4, Berlin, 1969.
[b] From Publ. No 6 (Canola Oil), Canola Council of Canada, Winnipeg, Man.
[c] 18:1 and 18:2 combined.

ual dietary sources of lipid (Table 5-5). The difference between the individual fatty acids is in the number of carbon atoms included in the chain of the molecule as well as in the type of bond between the carbon atoms (Table 5-5 and Figure 5-8). When some of the neighboring carbons are attached to each other in the chain by a double bond, we have an *unsaturated* fatty acid. Saturated acids have no double bonds in the chain. *Polyunsaturated* acids are those having more than one double bond between carbons in the molecule. The predominant types of fatty acids present in any particular fat or oil will determine its behavior in regard to temperature.

Fatty acids with short chains or a higher degree of unsaturation have lower melting points than long chain or saturated acids of corresponding chain lengths. Since most vegetable oils contain a higher proportion of the mono- or polyunsaturated acids, they tend to be liquid at room temperature, while the opposite is true for animal fats and their products (butter, lard). Technologically, this is the basis for the margarine manufacturing process (Section 5.8).

The much debated role of saturated vs. polyunsaturated fatty acids as related to human health and cardiovascular disease has been the subject of perhaps the most controversial scientific dispute in the history of food technology and human nutrition. Some of the particulars of this ongoing dilemma are further discussed in Section 5.10. The sharply antagonistic spirit is most unfortunate since it makes adversaries of the dairy industry as the manufacturer of butter, and the oilseed industry with its ingenious process of converting vegetable oil into margarine. Both of these highly successful industrial table spreads are safe and nutritious products of modern food industry and there is no need to tarnish their reputations by attacks and counterattacks based on unproven hypotheses.

Oil Extraction

The contemporary technology of oilseed processing is based on the well-known fact that fats and oils are not soluble in water but in organic solvents. Thus, if a crushed seed is mixed with a liquid organic solvent such as petroleum ether or hexane, the oil dissolves in the solvent, leaving most of the other seed components behind. The recovery of this *crude oil* by total evaporation of the solvent from the mixture completes the extraction cycle. Further refining is used to remove impurities contained in the crude oil. The expensive solvent (today almost universally hexane) is recovered by evaporation, condensed to the liquid state and used repeatedly.

A typical countercurrent oil extraction process is schematically shown in Figure 5-9. In practice, the seed is first steamed and crushed ("flaked") to rupture the cellular structure and expose the free oil to the action of the hexane. The flakes are then contacted with the solvent in a series of steps represented by compartments of the extraction vessel. The countercurrent movement of the hexane and the flakes may be accomplished by trickling the hexane through the extraction compartments so that the fresh flakes come into contact with the most concentrated oil–hexane solution. The fresh hexane is contacted with the most extracted flakes just before they are released. This arrangement is important for maximum recovery of the valuable oil. An efficient, economically viable operation should extract more than 99 percent of the oil content of the seed.

An older technology based on high mechanical pressure to expel the liquid oil from the seed without any external source of heat (*cold*

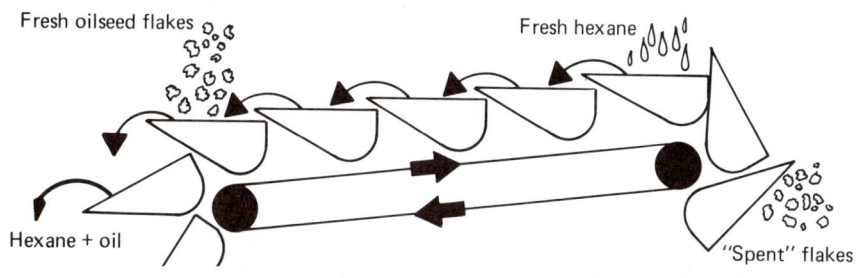

FIGURE 5-9. Schematic diagram of a countercurrent oil-extraction process with five extraction stages.

pressing), is still used as a preliminary step with oilseeds of very high oil content (typically, canola or sunflower). Sole use of cold pressing, as advocated by some authors who are critical of modern food-processing methods, would not be economically viable for an industrial process since at least 10 percent or more of the oil is not recoverable from the seeds by this *expeller* technology.

An experimental technique using water as the extraction medium is currently being evaluated by several research groups. The technique is based on emulsification (Section 5.9) rather than solvent extraction of the oil. The perceived advantage of this technique—apart from avoiding the use of the highly flammable, explosive and expensive hexane—is its potential for simultaneous recovery of good quality protein. The lower efficiency of the process in terms of the oil extracted is a major drawback, hindering its acceptance by industry.

Protein Meal Recovery

The oilseed residue remaining after the oil extraction, referred to as *spent flakes* or *meal*, is usually high in protein (up to 55 percent in the case of soybeans). Most of the meals are used as cattle feed. However, use of the oilseed meal protein for human food products has been steadily growing. Further processing of soybean meal into *protein concentrates* (characterized by the minimum protein content of 70 percent) or *isolates* (at least 90 percent protein content), is now an important industrial activity of several specialized soybean protein processors. These concentrated protein sources are used as raw material for various texturized vegetable protein products that may simulate meat, seafood, nuts, and other expensive sources of food protein. Even more important is the rapidly expanding use of soy protein ingredients with special functional properties such as heat-gelling, emulsifying, or water binding required in many traditional foods (Figure 5–10). Concerted research efforts of U.S. scientists in the area of soybean processing have resulted in rapid advances of the soy protein industry into many areas of food ingredient technology previously reserved for the more conventional suppliers typified by the dairy processor.

5.8 MARGARINE AND OTHER VEGETABLE OIL PRODUCTS

The main uses of vegetable oils are as raw materials in cooking, frying and formulation of many homemade or industrially processed foods, as salad oils, and in various further processed products such as short-

FIGURE 5-10. Examples of food products manufactured with soy protein ingredients. (Photo courtesy of Ralston Purina Co.)

ening, salad dressing or margarine. For shortening and margarine manufacture, the oils must be first *hydrogenated*, i.e. some of the unsaturated bonds in the molecules of the fatty acids must be chemically saturated by adding hydrogen atoms. This is necessary to control the product's plasticity, related to the "average" melting point of the various fatty acids present, and resulting in the highly desirable spreadability of refrigerated margarine. Precise control of the chemical hydrogenation process gives the margarine manufacturers a significant advantage over the butter processors, who are not permitted to modify the fatty acid content or the degree of saturation of the milkfat as produced by the cow. Thus, a wide spectrum of margarines is being manufactured, from highly unsaturated liquid products to cooking margarines, which are similar in saturation to butter.

In principle, the hydrogenation reaction is accomplished by bubbling or stirring hydrogen gas into a batch of pressurized hot oil containing finely powdered nickel. Once the process reaches the desired degree of hydrogenation, the nickel, which is present as a catalyst of this otherwise difficult reaction, must be removed by filtration. Hydrogenation results in production of trans-, as well as cis-fatty acids. These terms are related to the relative position of the hydrogen atoms added to the fatty acid molecule as shown in schematic representation of the hydrogenation reaction in Equation 5–1:

$$\begin{array}{c} \text{H H H H} \\ | \; | \; | \; | \\ -\text{C}=\text{C}\cdots\text{C}=\text{C}- \; + \; \text{H}_2 \end{array} \xrightarrow[\text{heat}]{\text{nickel}} \begin{array}{c} \text{H H} \quad\;\; \text{H} \\ | \; | \quad\;\; | \\ -\text{C}-\text{C}=\text{C}-\text{C}- \;\text{(trans)} \\ | \quad\quad | \; | \\ \text{H} \quad\;\;\; \text{H H} \\ \\ \text{H H H H} \\ | \; | \; | \; | \\ -\text{C}-\text{C}=\text{C}-\text{C}- \;\text{(cis)} \\ | \quad\quad\quad | \\ \text{H} \quad\quad\;\;\; \text{H} \end{array} \quad \text{(Equation 5–1)}$$

Trans-fatty acids are rarely present in naturally occurring vegetable fats and lipids, and it has been suggested that they may not be as nutritionally acceptable as the cis form. An alternative to hydrogenation, a process known as *interesterification*, is now being advocated as being more appropriate, although it is also based on the use of high heat (up to 200°C) and various catalysts. The main difference from hydrogenation is that no hydrogen atoms are added to the fat; rather the naturally present fatty acids are rearranged in their combinations with the glycerol molecule (Figure 5–8).

When hydrogenated vegetable oils, alone or in combination with other fats (lard or beef fat) are used without further processing we speak of shortening. For production of margarine, which originated (and is still perceived) as imitation butter, the hydrogenated vegetable fat is further mixed with a small amount of protein (usually skim milk powder) and water. This gives the final emulsion the water-in-oil character typical of butter. The differences between shortening and table margarine or butter are shown in Table 5–6. Various emulsifying agents, preservatives and other approved food additives (Chapter 9) are often added to optimize the sensory properties of this ingenious product of the modern food technology.

5.9 OIL EMULSIONS

Many consumer products made from lipids are mixtures of the fats or oils with water. A physically stable system containing two mutually insoluble liquid materials is called an *emulsion*. Water and oil are not soluble in each other. To form a stable emulsion, one of the components must be dispersed in the other component in the form of small droplets. Depending on which component forms the *dispersed* phase—as opposed to the *continuous phase* of the other component—the system is called oil-in-water (o/w) or water-in-oil (w/o) emulsion. Thus, certain salad dressings, mayonnaise, as well as milk, cream and some other dairy products, are o/w emulsions. Butter or margarine, which contain about 16 percent water in no less than 80 percent of the fat phase, are w/o emulsions.

The stability of an emulsion, indicated by the lack of visible forma-

Table 5–6 Compositional Data for Butter, Margarine, and Shortening

	TYPE OF FAT	% FAT	% WATER	% NON-FAT-SOLIDS[b]
Butter	Milk fat only	≥ 80	≤ 16	3–5
Margarine	Hydrogenated vegetable oil	≥ 80	≤ 16	3–5
Shortening	Mixture of hydrogenated vegetable oils	100	0	0

COMPOSITION[a]

[a] Based on "Nutrient value of some common foods", Publ. No. H58–28/1977, Health Protection Branch, Ottawa, Ont, and legally defined standards of identity.
[b] Includes added salt.

tion of big droplets or two separate layers, depends on the *size* of the dispersed globules and on the prevention of coalescence of small globules into progressively larger ones. There are fairly large differences in densities of water and oil (1.0 g/cm³ *vs* 0.9 g/cm³, resp.). In an o/w emulsion, the lighter phase (oil) has a tendency to rise, forming a "cream layer" on top. If the globules of the dispersed phase are sufficiently small, the velocity of their movement will be negligible. Theoretically, the creaming velocity of the globular phase follows Stokes's law of sedimentation of a solid sphere in a liquid medium (Equation 5-2):

$$v = \frac{d^2(S_g - S_l) \cdot g}{18\mu} \qquad \text{(Equation 5-2)}$$

It is evident that the linear creaming velocity v (its units are generally distance/time) is directly and significantly influenced by the diameter of the droplet *d*, and also by the density differential between the dispersed globule (S_g) and the continuous liquid phase (S_l), and the gravitational acceleration *g*; the viscosity of the continuous phase μ is indirectly proportional to the creaming velocity.

To minimize the creaming of the dispersed phase, and to prevent formation of large droplets from the adjoining globules by their coalescence, it is necessary that a third compound be present in a stable emulsion. There is a class of naturally occurring or industrially produced compounds known as surface-active agents (or *surfactants emulsifiers*). Generally, molecules of surface-active agents have *hydrophilic* ("water-loving") heads and *hydrophobic* ("water-hating", also oil-loving, *lipophilic*) tails, as shown schematically in Figure 5-11a. When a small amount (typically 0.5% or less) of a surfactant is present in an emulsion, the surface-active molecules attach themselves to the dispersed globules so that the hydrophilic heads face the water phase and the hydrophobic tails the oil phase (Figure 5-11b). In a properly stabilized emulsion, the surfactant coating of the dispersed phase will prevent the droplets from coalescence upon impact; it may also diminish the density differential between the dispersed and the continuous phase (S_g and S_l in Equation 5-2). Other necessary properties of a stable emulsion are the small average size of dispersed globules (achieved by homogenization, as explained in Section 6.4) and, if possible, increased viscosity of the continuous phase μ. Margarine and butter are stable w/o emulsions because the continuous phase (the fat) has a high viscosity at the ambient temperature; upon heating, the fat will melt and the two phases may be separated.

Many emulsifiers are naturally present in certain food systems. Examples of naturally occurring surfactants include phospholipids in milk or lecithin in soybeans. Industrial emulsifiers, mostly derivatives

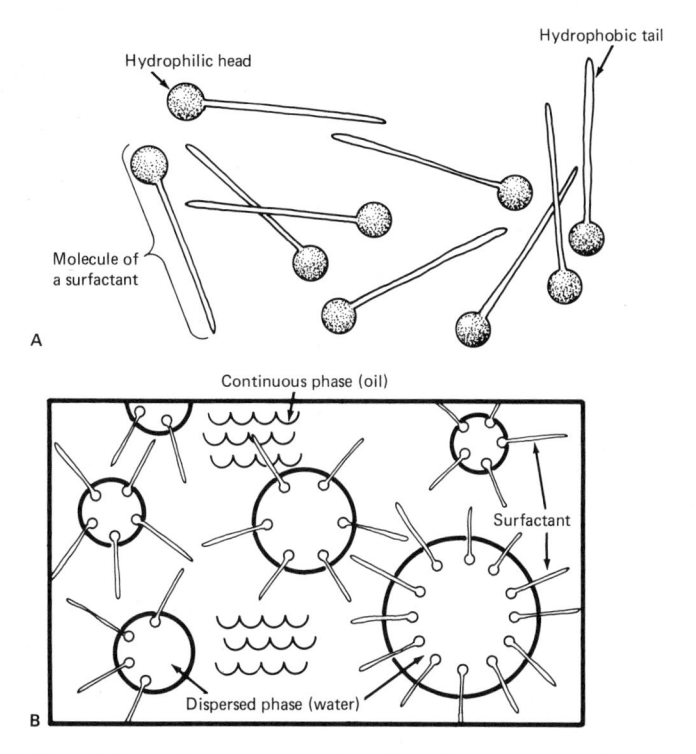

FIGURE 5-11. Schematic representation of surfactant molecules (A) and their distribution in an in-water oil emulsion (B).

of certain fatty acids, are manufactured from natural sources or chemically synthesized in a laboratory. Many of the present oil-based foods could not be produced without the use of these natural or man-made food additives.

5.10 NUTRITIONAL ASPECTS OF CEREAL AND OILSEED PRODUCTS

Despite their unquestionable importance as sources of dietary energy, cereal and oilseed commodity groups appear to be the focus of more nutritional controversies than all other food materials combined. The facts and fancies of some of the controversial topics were briefly exposed in the preceding paragraphs. These include perceived nutritional advantages or health hazards of stone grinding and bleaching of flours; the myths and virtues of bran and other fiber sources; the criticism of breakfast cereals; the cold pressing of vegetable oils; the ongoing trade war between butter and margarine; and the underlying emotional rather than rational issue of polyunsaturated vs. saturated fat. Additional related topics of interest that are continuously appearing in scientific and popular literature include the comparison between vegetable and animal sources of fat regarding cholesterol, the marketing of emulsifiers (especially lecithin) in tablets as a health food item, or the magical properties of megadoses of fat soluble vitamins A and E.

The main aspect of cereals and oilseeds in human nutrition is not related to any of these controversial issues, processing techniques or unproven positive or negative dietary effects. Rather, both of these commodities provide food components that are valuable sources of food energy, which in oversupply may lead to health complications. The additional calories consumed in excess of what our bodies need have been called by a reputable medical journal the most dangerous of all "food additives" in terms of human health. The severe medical complications related to obesity are recorded in health statistics both in the U.S. and Canada.

Of the two principal components of cereals and oilseeds, fat furnishes more than twice as much food energy as starches and related carbohydrates. There is "invisible fat" in many common items on our menu, such as beef steak, cheddar cheese, or party peanuts. Additional fat is absorbed during the manufacture of popular foods such as potato chips or French fries, doughnuts or cookies from vegetable oils used in their processing. One of the few generally agreed upon dietary guidelines for North Americans is to decrease the total fat intake to not more than 20 to 30 percent of total calories (today's typical figure is about 40 percent). About 65 percent of our dietary calories should be supplied by carbohydrates; this can be accomplished by increasing consumption of cereal foods.

While decreasing the overall consumption of fats appears desirable, the recommendations regarding the kind of fat (vegetable oils vs. animal fat) are highly controversial. In recent years, theories, experimental data and epidemiological studies were used with increasing frequency to support—or counterattack—the notion that polyunsaturated fats (coming from vegetable oils) are beneficial for human health, particularly as related to cardiovascular disease. There are countless studies that tend to prove or disprove almost anything in this highly emotional debate. Some of the studies show clearly that diets containing highly unsaturated oils can cause severe heart lesions in experimental animals. The food use of canola oil is still not allowed in the U.S. unless the oil is completely hydrogenated, since the high content of erucic acid in the older varieties of rapeseed has been linked to coronary hazards. (Erucic acid is unsaturated.) On the contrary, the leading proponents of polyunsaturated fat quote many studies and epidemiological data showing lower incidence of coronary heart disease among populations consuming predominantly fish or vegetable lipids as opposed to fat from red meats or butter. Some of these studies surveyed fishermen and other groups of individuals whose diet (but also lifestyle) was quite different from the typical urban dweller facing the daily pressures of many stressful situations. Because of the highly individual metabolic response of a human body to dietary fat intakes and the large variations in natural blood cholesterol levels (one of the proven indicators of a coronary risk), it is doubtful that this controversial issue will be resolved in the near future. As a major long-term nutritional study showed in the U.S. in 1984, controlled exclusion of cholesterol-containing foods from diets of individuals with high levels of blood cholesterol is almost totally ineffective in lowering the blood cholesterol —or reducing the high risk of heart attack for these individuals. The relationship between dietary fat intake and blood cholesterol is equally inconclusive, although decreasing the total fat consumption is generally recommended as desirable for reduction of the coronary heart disease problem.

In terms of specific requirements for macronutrients supplied by oils and cereal products, there is only one essential component— linoleic acid. The minimum recommended dietary allowance for this unsaturated fatty acid is low (about 2 percent of total calories) and is easily supplied by any adequate diet so that nutritional deficiencies are virtually unknown. There are no specific nutritional requirements for starch or fiber components of cereals, although in general, both are desirable dietary compounds and both are amply supplied by the cereal-based foods. Recommendations for increased consumption of dietary fiber have appeared relatively recently as a possible, but unproven, prevention of diverticulosis, ischemic heart disease and cancer of the colon. Cereal products made from whole flours are a good source of dietary fiber, and many breakfast cereals have extra fiber added. However, cereal fiber has no magic curative properties and many other foods, particularly fruits and vegetables such as corn, peas, beans or apples, are just as valuable sources of dietary fiber as the revered bran. The nutritional inferiority of plant proteins, due to lack of certain essential amino acids (Table 1-10), is well known and may be significant for strict vegetarians. Should the use of fabricated plant proteins increase substantially in the future at the expense of traditional animal protein sources (meat, fish, dairy products), amino acid deficiencies could become more worrisome for general populations. At the present time, however, cereal and oilseed products should be enjoyed in moderation as valuable sources of food energy, and as many truly special and tempting food items typified by freshly baked bread, birthday cake with rich frosting, or shrimp cocktail with mayonnaise.

5.11 CONTROL QUESTIONS

1. Compare the technological requirements for storage of fruits and vegetables, and cereals and oilseeds. What is the main difference between these two groups of plant products?
2. What are the purposes of contacting grain kernels with water before milling and malting? What are the approximate moisture contents of the conditioned wheat before milling and that of barley before drying the germinated malt?
3. Is it a good practice to store packaged bread in a refrigerator? Why or why not?
4. In a mass balance check of a milling operation producing straight flour it was found that in an eight-hour operation there were 240 bags (weighing 50 kg each) produced. The amount of bran obtained during this operation was 3,200 kg. The flour analysis showed that the flour was of extraction typical for straight flour. How efficient was the milling operation in terms of flour losses, if the capacity of the processing line was rated at 2,000 kg of wheat milled per hr?
5. What is the function of hexane in a countercurrent extraction of oil from soybeans? What are the technological differences between solvent extraction and cold pressing? Are there any nutritional differences in the two types of products (extracted oils) obtained?
6. What is hydrogenation and what major product is based on this chemical reaction? Why does rapeseed oil have to be fully hydrogenated before being allowed entry into the U.S. food market?
7. Explain the terms "w/o emulsion," "surfactant," "continuous phase," and "creaming."

**Suggested reference books —
Cereals and Oilseeds**

Hummel, C. 1966. *Macaroni Products*. London: Food Trade Press.
Jenkins, S. 1975. *Bakery Technology*. Toronto, Ont.: Lester and Orpen, Ltd.
Pomeranz, Y. and Shellenberger, J. A. 1971. *Bread Science and Technology*. Westport, Conn.: AVI Publ. Co.
Pomeranz, Y. 1978. *Wheat - Chemistry and Technology*. St. Paul, Minn.: Amer. Ass'n. of Cereal Chemists.
Pyler, E. J. 1982. *Baking Science and Technology*. Chicago, Ill.: Siebel Publ. Co.
Smith, A. K. and Circle, S. J. 1978. *Soybeans: Chemistry and Technology*. Westport, Conn.: AVI Publ. Co.

Spiller, G. A. and McPherson - Kay, R. 1980. *Medical Aspects of Dietary Fiber.* New York: Plenum Publishing Co.

Vaisey - Genser, M., and Eskin, M. 1982. *Canola Oil - Properties and Performance.* Winnipeg, Manitoba: Canola Council of Canada.

Weiss, T. J. 1983. *Food Oils and Their Uses.* Westport, Conn.: AVI Publ. Co.

6

Milk and Dairy Products

6.1 MILK PRODUCTION

Milk is often called the near-perfect food. This is not surprising since milk was intended by Mother Nature to be the sole food of newborn mammals, including humans. Throughout the world, cow's milk is the most predominant raw material for the diverse products of the modern dairy industry. Milk of other mammals (goat, sheep, buffalo, even reindeer) is being used to a limited extent in some countries and by some ethnic groups.

The production of cow's milk is a typical farming operation, starting with the birth of a calf. The lactation period for a cow lasts about 43 weeks and all lactating cows in a herd are milked twice daily. Figure 6-1 shows one of the several possible arrangements of a milking parlor. Immediately after milking, the fresh milk is cooled to 3–5°C in refrigerated farm tanks where it is stored until collected by the dairy processor, usually every one or two days. Modern milk collection systems are based on bulk tank trucks (Figure 6-2), which call on several farms along each pickup route, delivering the mixed milk to the dairy processing plant in bulk. In North America the milk can is now virtually extinct, although in many other countries mixed collection systems are still used, combining the ease of can delivery from remote farms with efficient road-tanker service.

Refrigerated on-farm milk storage is one of the major advantages of bulk collection systems. Rapid cooling of fresh milk is very difficult with milk cans. Cold storage and better sanitation procedures used with farm storage tanks and other modern milking equipment are cru-

FIGURE 6-1. A milking parlor of a dairy farm.

FIGURE 6-2. A bulk tank truck used to transport milk from farms to a dairy factory. (Courtesy of Palm Dairies Ltd.)

cial for low microbial content of raw milk. Good quality of a finished dairy product cannot be achieved when bacteriological quality of the raw milk is poor. Dairy farmers consistently producing milk with low bacterial count are sometimes rewarded with bonus payments. The principal considerations that determine the farm value of milk that is produced are its volume and fat content, and in certain payment schemes also the protein content.

6.2 MILK COMPOSITION

In this text, as in food regulations of many countries, "milk" without any further indication denotes cow's milk. Milk can be defined as being the normal lacteal secretion obtained from the mammary gland of the cow. Besides water, normal cow's milk contains about 5% of milk sugar called *lactose*, about 3.2% *milk protein* consisting of several chemically different proteins, 3–5% or more *milk fat* depending on breed, type of feed, season, and other factors (a typical value for bulk delivered milk is 3.6–3.8% fat), and about 0.75% minerals. Table 1–3 gives the approximate composition of cow's milk in comparison to human and goat's milk. The principal constituents of these proximate component groups are listed in Table 6–1. Before processing, the total solids content of the fluid milk is about 13%, which is more than in many solid foods, especially fruits and vegetables.

In the physicochemical sense, milk is a very interesting system simultaneously representing a true *solution* (salts and lactose in water), *emulsion* (milk fat globules in the water solution), as well as *colloidal* dispersion (protein particles). Casein, the major milk protein, is not truly soluble in water; in the milk system it is dispersed in small units called *casein micelles* which do not settle under normal gravitational conditions. This gives milk its whiteness, since the casein micelles, containing several hundred casein molecules, are large enough to scatter light.

In addition to the proteins, lactose, fat and minerals, milk contains a wide variety of minor components such as vitamins, enzymes, trace elements and other compounds. Some of these play an important role in dairy processing and/or in human nutrition. Table 6–2 gives a partial list of some of the minor constituents significant for the dairy industry.

6.3 FLUID MILK PROCESSING

Dairy processing plants can be roughly classified into two types—fluid milk plants and manufacturing plants. The *fluid milk plants* are situated close to the market (in big cities) and their products include

Table 6-1 Principal Constituents of Cow's Milk (Average Values)

COMPONENT GROUP (CONTENT IN MILK, % w/w)	MAIN COMPONENTS	PROPORTION OF COMPONENTS WITHIN COMPONENT GROUP %
Proteins (3.2):		
	Casein	80
	Whey proteins:	15, of which
	α-lactalbumin	3
	β-lactoglobulin	9
	Other minor proteins	3
	Non-protein nitrogen	5
Fats (3.7):		
	Milk fat	≥ 98
	Fat globule membrane	≤ 2
Carbohydrates (4.9):		
	Lactose	>99
	Glucose	<1
	Galactose	Trace
Minerals and Salt Constituents (0.75):		
	Calcium	18
	Sodium	8
	Potassium	20
	Phosphorus	13
	Chlorine	16
	Other (magnesium, citrate, carbonate)	25

various types of pasteurized fluid milk, numerous fermented products such as yogurt, cottage cheese, buttermilk, sour cream, kefir and others, often ice cream and sometimes butter. The *manufacturing plants* usually process large quantities of milk into various less perishable products such as cheese, evaporated and dried milk products, and various other industrial products for food or non-food uses (casein, lactose, whey protein concentrate and many special products). Butter or ice cream may also be manufactured in some of these plants, frequently located close to major milk producing areas.

There are several operations in the primary treatment of raw milk common to all plants. The raw milk storage area is typically equipped with large vertical *silo tanks* with capacity often exceeding 100,000 liters (Figure 6-3). The raw milk is pumped into these huge silos from several road tankers as soon as they arrive from the collection routes.

Table 6-2 Minor Constituents of Cow's Milk With Importance to Dairy Industry

COMPONENT GROUP	EXAMPLE	IMPORTANCE TO DAIRY INDUSTRY
Enzymes	Lipase	Fat breakdown (rancidity) in improperly processed products
	Alkaline phosphatase	Indicator of proper pasteurization treatment
	Protease (also of microbial origin)	Bitterness in cheese and other products
Vitamins	Vitamin A + carotenoids	Nutritional importance (good source); yellow color of butterfat
	Riboflavin (Vitamin B_2)	Nutritional importance (excellent source); light sensitive; greenish color of cheese whey
	Vitamin D	Nutritional importance for dietary absorption of calcium; added since milk contains little
Trace elements	Copper	Catalyzes undesirable oxidation of milk fat; some nutritional importance
	Zinc	Nutritional importance

This system of raw milk handling requires the cooperation of milk producers.

It is imperative that only good quality raw milk is offered for shipment from the farm. One careless farmer could spoil the whole content of the factory silo tanks, by including in his shipment milk that contains antibiotics used for treatment of cows suffering from mastitis, an inflammatory disease of the udder. Most processing plants today employ routine checks for antibiotics on all incoming raw milk before it is pumped into the silo tanks.

Virtually all raw milk is pasteurized before it is further processed into the various final products. Pasteurization is similar to blanching of vegetables in that it employs controlled heating of the milk. While in blanching of vegetables the primary task is the deactivation of enzymes, the most important function of pasteurization is to eliminate all *pathogenic* microorganisms that may be carried in the milk from the cow, particularly the undulant fever, tuberculosis, Q-fever and other diseases transmittable to humans. The particular requirements for

130 Milk and Dairy Products

FIGURE 6-3. Storage tanks for raw milk in a dairy factory.

pasteurization are specifically designed to kill *Coxiella burnetii*, the most resistant pathogenic organism commonly associated with cow's milk. The regular pasteurization process used today consists of heating the milk to not less than 72°C, holding at this temperature for 15 sec., and cooling rapidly to minimize the cooked flavor development typical of severe heating. This treatment, usually referred to as High-Temperature-Short-Time pasteurization or simply, HTST, also greatly reduces (although not fully eliminates) the microbial populations causing spoilage. In addition, pasteurization also deactivates most enzymes present in raw milk that would cause undesirable biochemical changes in the finished product, especially rancidity (enzymatic breakdown of fat), or bitterness, caused by breakdown of protein. The destruction of the enzyme *phosphatase*, which is slightly more heat resistant than the most heat resistant pathogens and which is easy to test for, is used as an indicator of adequate pasteurization treatment.

The HTST *pasteurizer* (Figure 6-4) is a *heat exchanger* consisting of a multitude of stainless steel plates held in a frame. The milk is pumped through the machine so that it flows on one side of the thin, corrugated plates; the heating or cooling medium flows on the other side. The cold, raw milk enters the *regeneration* section of the pasteurizer where it

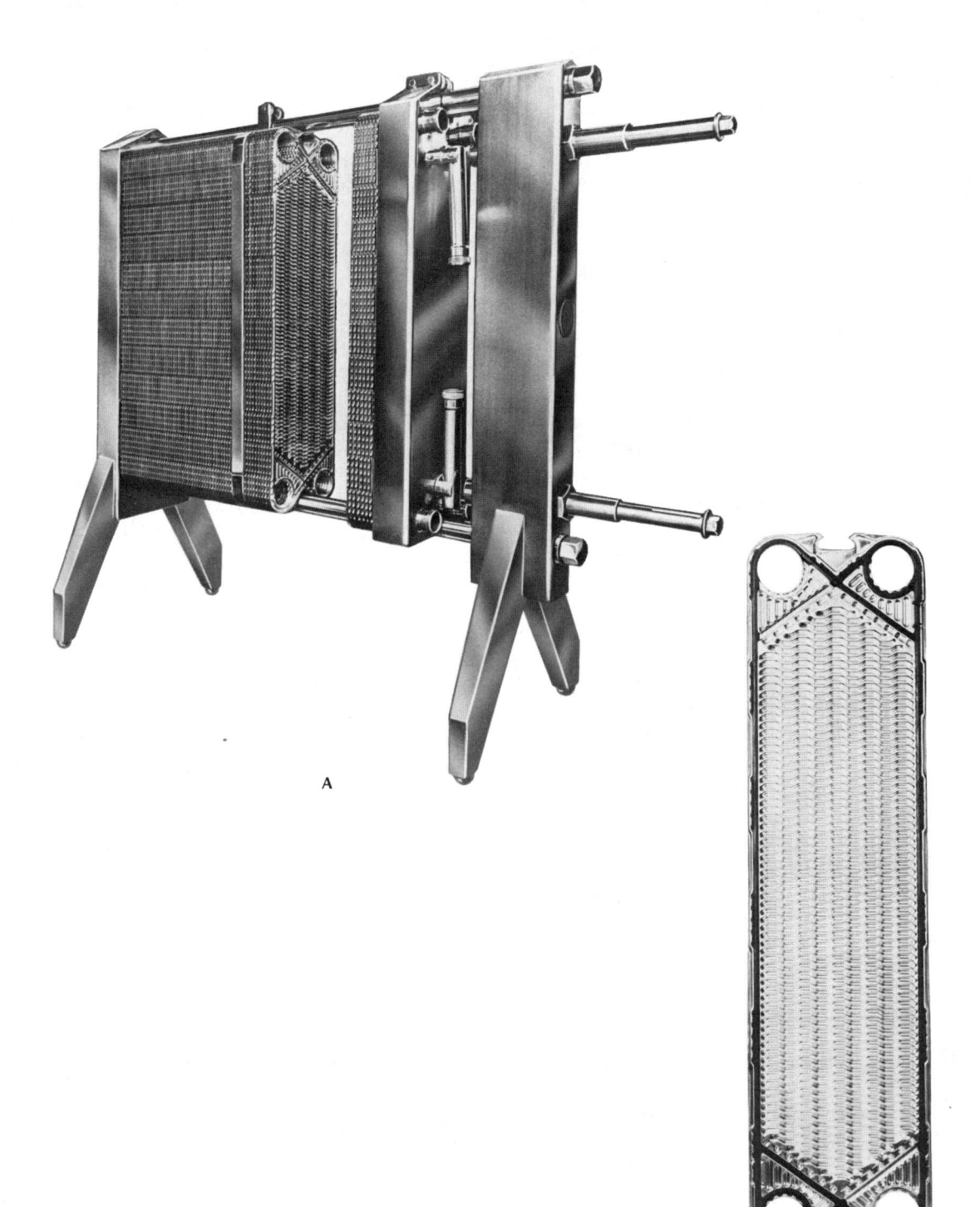

FIGURE 6–4. The HTST milk-pasteurizer (A) and its main component, the corrugated stainless steel plate (B). (Courtesy of APV-Crepaco Canada)

is used as a cooling medium to cool the hot pasteurized milk. This transfers heat back to the incoming raw milk, warming it up and saving precious energy. In the final heating or cooling sections, either hot or cold water is used. There is no mixing of the milk with the heating or cooling media at any point in the HTST pasteurizer; this is insured by proper design safeguards including heavy rubber gaskets separating the plates.

Most of the pasteurized milk destined for fluid milk products is further treated by *separation* and/or *homogenization*. Both of these processing operations are concerned with milk fat. When a fat-reduced product such as skim milk—or partially skimmed milk usually containing 2% fat—is desired, the excess fat is removed from the normal milk by centrifugal separation.

The fat globules in the normal milk are approximately 4 μm or more in diameter. Since the fat is lighter than the aqueous phase, these large droplets will migrate slowly to the top if the milk is left undisturbed, eventually (in several hours) forming a thick layer of cream. In a centrifugal dairy separator, this creaming process is accomplished in a matter of seconds, the cream portion of the milk being continuously removed from the skim milk portion.

In order to avoid undesirable creaming in fluid milk products containing fat, the milk is *homogenized* by pumping through a special valve under high pressure. This purely mechanical treatment breaks the large fat globules into a multitude of very small ones, on the average below 1μm in diameter, which is sufficiently small to prevent the undesirable creaming. In the production of partially skimmed milk and other products with well-defined fat content, the *standardization* of the final desired fat content is accomplished by mixing appropriate amounts of homogenized and skimmed milk either in a mixing tank after complete separation, or in a special centrifuge during the separation process.

Excess milk fat obtained by the separation of cream is used as a raw material for coffee cream, whipping cream, and other dairy cream products listed in Table 6-3. Cream products differ mainly in the fat content and are produced by standardization of cream separated from milk by further centrifugation or by addition of skim milk. All creams are still o/w emulsions since the milk fat globules are well protected from coalescence by the natural emulsifiers present in milk.

Packaging of the processed fluid milk products is a very important operation with respect to the final product quality. Any microbial contamination during this step (often described as post-processing contamination since it occurs after the pasteurization treatment) would result in poor shelf life or even health hazards. Appropriate packaging is also important to preserve some of the light-sensitive milk vitamins,

Table 6-3 Protein, Fat, and Water Contents[a] of Some Fluid and Fermented Dairy Products (orientation values)

PRODUCT	% FAT	%PROTEIN	%WATER
Homogenized milk	3.5	3.5	87.5
Skim milk	0.1	3.6	90.5
"Half-and-half" (cream and milk)	12.0	3.2	79.7
Coffee cream	20.0	3.0	71.5
Whipping cream	37.0	2.2	56.6
Sour cream	14.0	2.8	76
Buttermilk	0.1–2.0	3.6	90.5
Yogurt	2.0	4.5	85.0

[a] Data from "Nutritive Value of American Foods," by C. F. Adams, Agric. Handbook No. 456, USDA, Washington, D.C., and other sources.

particularly riboflavin. For this reason, a multilayered, laminated paper carton or a light-impermeable plastic bottle is preferable as milk packaging material to glass or a clear plastic pouch. Properly packaged fluid milk stored in a refrigerator can be kept for up to three weeks without much deterioration in quality. Such prolonged shelf-life is common in the North American dairy industry, while the refrigerated shelf life of regular pasteurized milk in most European countries rarely exceeds a few days. This enviable accomplishment of modern dairy technology is due to the advancements in process engineering, sanitation and quality control, applied in the plant and especially on the farm. No chemical preservatives are permitted in any fluid milk products.

6.4 CULTURED DAIRY PRODUCTS

The technology of unripened cultured milk foods, such as buttermilk, sour cream, yogurt, cottage cheese, quarg, kefir, or acidophilus milk, is based on microbial conversion of the milk sugar, lactose, to lactic acid. This *lactic fermentation* process is accomplished by inoculating special strains of selected microorganisms (*lactic culture*) to pasteurized milk. There are many different lactic bacteria which have specific *incubation temperature* requirements and which, besides the lactic acid, also produce the many characteristic delicate flavor compounds typical of fermented milk products. For example, sour cream is made by fermenting dairy cream with bacteria of the Streptococcus spp. that usually grow well at room temperature and produce minute quantities of the desired flavor compound, diacetyl.

A similar fermentation process is used in the production of buttermilk. In North America this usually means cultured skim or partially skimmed milk. Originally, buttermilk was a byproduct of a buttermaking process which used fermented cream for the buttermaking. Nowadays, most butter is produced from unripened ("sweet") cream and sweet buttermilk is not utilized in this form.

Yogurt must be incubated at elevated temperatures (45–50°C); the two types of microorganism used as a yogurt culture (*Lactobacillus bulgaricus* and *Streptococcus thermophilus*) are thermophilic and would not grow at room temperature. The yogurt cultures produce the compound acetaldehyde, which gives natural yogurt its typical flavor resembling green apples and thus sometimes described as "green."

Acidophilus milk and kefir are sour milk products that are becoming increasingly popular. *Lactobacillus acidophilus* is used in production of acidophilus milk for its alleged therapeutic properties. In kefir fermentation the kefir grain biomass includes yeasts which give the final product its sparkling character caused by CO_2 and a small amount of alcohol.

Cultured dairy products are made from milk pasteurized before the fermentation. As a result, these products normally contain the viable lactic bacteria used in their manufacture. Most of these products can be prepared at home by inoculating pasteurized milk with a spoonful or two of the desired product purchased commercially, and keeping it at the required temperature for several hours. Optimum temperatures for production of some fermented dairy products are listed in Table 6-4. Only fresh products should be used for inoculation, and contamination with undesirable bacteria must be avoided. For yogurt production milk should be enriched with about 2–3% non-fat dry milk and the mixture heated to about 90°C for at least 5 min. This treatment will cause the whey proteins of milk to combine with the casein micelles. The resulting complex protein structures will entrap the surrounding water thus forming a smooth gel-like structure of the fermented product. In North America gelatin and other texture-improving additives are sometimes used to assist in the formation of this delicate consistency.

Although cottage cheese is included in the group of unripened cultured milk foods, it differs from the various sour milk products in one important aspect. The lactic acid fermentation is used not only for flavor development, but also for "curdling" of the casein which is subsequently separated from the rest of the original milk. In this the cottage cheese technology combines the principles of cultured dairy products' manufacture with the fundamental aspect of cheesemaking—the separation of protein coagulum from the rest of the milk components. For cottage cheese manufacture, casein is *coagulated* and *precipitated*

Table 6-4 Major Types of Lactic Bacteria Used in Fermented Dairy Foods

PRODUCT	TYPICAL CULTURES USED	OPTIMUM TEMPERATURE (°C)
Cultured buttermilk and sour cream	*Streptococcus lactis* or *S. cremoris* and *Leuconostoc citrovorum* or *S. diacetylactis*	22–25
Kefir	Kefir grains (containing *Sacharomyces kefir, Torula kefir, Lactobacillus* and *Leuconostoc* spp.)	18–22
Yogurt	*Lactobacillus bulgaricus* and *Streptococcus thermophilus*	45–47
Cottage cheese or quarg	*S. lactis* and/or *S. cremoris* and *L. citrovorum* and/or *S. diacetylactis*	32 (short set)[a] or 22 (long set)[b]
Acidophilus milk (fermented)	*Lactobacillus acidophilus*	38

[a] 5–6 hrs
[b] Overnight (14–16 hrs)

from its colloidal state when the acidity is increased, or more precisely, when enough lactic acid is produced for the system to reach about pH 4.8. The remaining components of the original *skim milk* from which the cottage cheese is made (i.e. most of the lactose, the acid-soluble whey proteins, the minerals including the valuable calcium, and water-soluble vitamins), are discarded in whey. The whey disposal is one of the most pressing food-related pollution problems, since up to 9 kg of whey is generated for every kg of cheese made.

To improve the structure of the cottage cheese grain, a small amount of a milk-clotting enzyme called *rennet* is used similarly as in the manufacture of other cheeses explained below. Texture is also strengthened by heating ("cooking") the mixture of curds and whey at 45–50°C. Without the rennet addition and the cooking step, the lactic acid fermentation of milk will produce a fragile, pasty casein curd that can be separated by straining through the filters or cloth, or by centrifuging. This process is used to produce baker's cheese or quarg, a soft unripened "cheese paste" popular in Germany and other European countries. The cottage cheese making process is schematically shown in Figure 6-5 in comparison to the cheddar cheese making, and the cottage cheese composition is compared to that of other cheeses in Table 6-7.

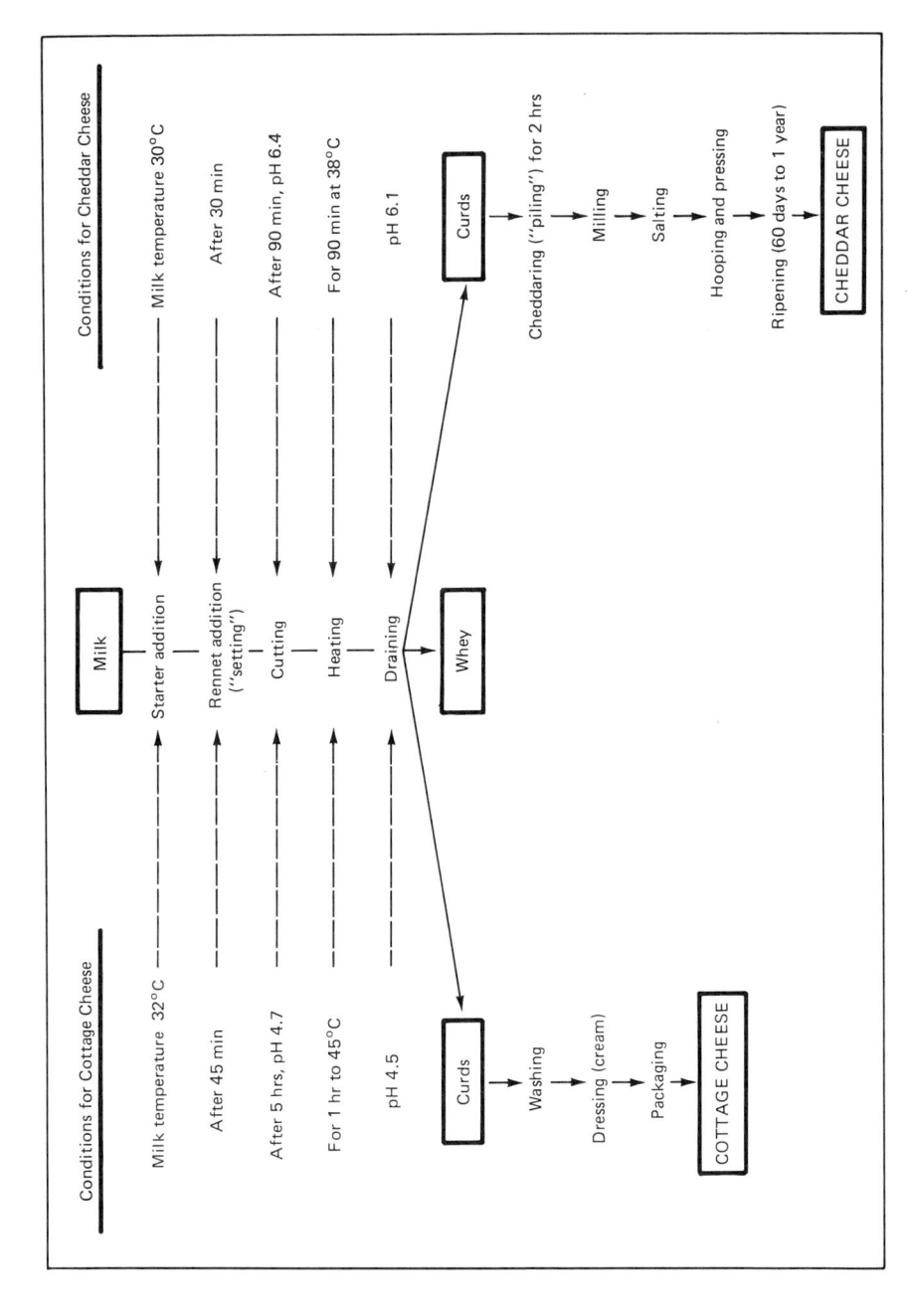

FIGURE 6-5. Schematic diagram of manufacturing conditions for cottage and cheddar cheese-making.

Because of its high moisture content and casein coagulation by low pH—at which the milk calcium is fully soluble in water and is thus lost in the whey—the cottage cheese has, pound for pound, a lower nutritional value than the "hard cheeses." Its value is as a principal source of balanced dietary protein with little fat, and thus is one of the mainstays of slimming diets.

6.5 CHEESE

Cheesemaking belongs among the most traditional food manufacturing processes used throughout the world. Although the technological principles are the same for most cheesemaking procedures, there are countless variations in the individual steps of the process which account for all the cheese varieties (several hundreds of them) that are being produced from cow's, sheep's, goat's and other milks. Table 6–5 gives one common approach used to classify cheese types, based primarily on the amount of moisture retained in the cheese. While science made significant progress in the understanding of the whole cheese-making process, there are many instances where the old saying that "cheesemaking is more an art than a science" still applies.

Table 6–5   Main Classes of Natural Cheese

CONSISTENCY	EXAMPLES	APPROXIMATE H_2O CONTENT[c] (% w/w)
Soft[a]	Cottage cheese	80
	Quarg	79
	Baker's cheese	74
	Neufchatel	62
Soft[b]	Camembert, Brie	52
	Limburger	50
Semi-soft[b]	Roquefort, Blue, Stilton	44
	Brick	42
	Munster	43
	Mozzarella	52
Hard[b]	Cheddar	38
	Gouda	41
	Swiss (Emmenthaler)	39
	Gruyere	34
Very Hard (Grating)[b]	Parmesan	31
	Romano	25
	Sbrinz	18

[a]Unripened
[b]Ripened
[c]Average values obtained from various sources

The principal steps of a cheesemaking sequence, shown in Figure 6-5, fulfill various technological functions. Although the variations in these steps used to produce the different cheeses are many, their principles are straightforward and common to all processes.

Starter and Rennet Addition

Similar to the fermented milks, "starter" cultures are primarily lactic acid bacteria used to produce acidity and other desirable flavor compounds by fermentation of the lactose. The microbial action, however, is not limited to the relatively short time of milk fermentation, and is *not* relied on to curdle the protein. In addition to lactic starter cultures, a great variety of other microorganisms (including various species of bacteria, as well as molds and yeasts) are added for secondary fermentation during the ripening stage. Table 6-6 lists several typical microorganisms and explains their role in producing certain cheeses.

The enzyme rennin, found in calves' stomachs, coagulates milk without the decrease in pH. This phenomenon has become the basis of the traditional cheesemaking process. Rennet, the industrial preparation of rennin, is used to coagulate the milk in the cheese vat approximately 30-60 min. after the starter addition. The milk turns into a smooth, elastic gel about 20-30 min after adding the rennet (approximately 1 g rennet is used for every 5 kg of milk). Rennet is produced by specialized supply companies from the stomachs of slaughtered calves, or from substitute microbial sources as calves are becoming scarce.

Cutting and Cooking

After the content of the cheese vat has "set" into the uniform rennet gel, special curd-cutting knives are used to break the gel into small cubes. This initiates the expulsion of whey from the gel. As a result, the vat content becomes a heterogeneous mass of curd cubes (comprised of protein and milk fat) floating in the whey. Raising the temperature of the vat content to about 35-50°C, according to type of cheese being made, will increase the firmness of the curds and will enhance the fermentation action of the starter organisms. Production of acidity in the curd continues during these and subsequent steps.

Draining, Matting and Washing

The "cooked" curd must be separated from the whey, which is accomplished by draining the whey from the vat through a sieve-like strainer. Typical pH of the rennet whey upon draining is about 6.0-6.2. Consequently, much of the calcium is not completely soluble in the whey at this pH and will be retained in the cheese (Table 6-7).

Table 6-6 Effects of Certain Microorganisms in Cheesemaking

CULTURE	TYPE OF ORGANISM	EFFECT
Lactic acid bacteria (*S. lactis, S. cremoris, L. casei, L. helveticus*, etc.)	Bacteria	Produce acidity during cheesemaking and remove residual lactose in early stages of cheese ripening; also produce proteolytic enzymes for protein structure modification.
Propionibacterium shermanii	Bacterium	Produces propionic acid and CO_2 during ripening of Swiss cheese resulting in the "eyes", and the sweet, "nutty" flavor.
Penicilium camemberti	Mold	Develops the "velvety" outer rind on Camembert-type cheeses; also breaks down the protein structure resulting in the creamy texture.
Penicilium roqueforti	Mold	Creates the blue veins in the Roquefort, Danablue, Stilton, and similar cheeses, catalyzes the breakdown of fat resulting in the characteristic "rancid" flavor.
Candida spp., *Mycoderma* spp. and	Yeasts	Used in surface ripening phenomena (color, flavor, surface shine) characteristic of Brick, Limburger, Oka and other cheeses.
Brevibacterium linens	Bacteria	

Substantial variations of the draining and subsequent curd-handling procedures exist in the processing technologies for the individual cheese types. In "Cheddar-making," the curd is not washed but is allowed to "matt" forming large blocks; this process, called "Cheddaring," is unique to Cheddar cheese. Curds for other cheeses may be washed, stirred in the whey and/or wash water, or drained without washing or matting.

Milling and Salting

For proper flavor, some control of the ripening process, and further whey expulsion, all cheeses are salted. The amount of salt used and the mechanics of the salting process add another important variable differ-

Table 6-7 Content of Selected Nutrients[a] Per 100 g of Some Cheese Products

	H_2O	PROTEIN	FAT	CALCIUM (mg)
	----------g----------			
Cottage cheese	79	15	4	60
Camembert	52	17.5	25	200
Blue (Roquefort)	44	21	30	350
Brick	41	22	30	700
Swiss (Emmenthaler)	39	28	28	925
Cheddar	37	25	32	750
Processed cheese	40	23	30	620

[a]From "Nutritive value of American foods," by C. F. Adams, Agric. Handbook No. 456, USDA, Washington, D.C., and other sources.

entiating the cheese varieties. In manufacturing of Cheddar, the matted blocks are milled into small slabs or cubes and granular salt is mixed into the "pile." Curd destined for Gouda cheese is first pressed into the proper final cheese shape, and the loaves are immersed into a salt brine. Colby and other cheeses are salted during the curd washing process.

Forming and Pressing

To form the final cheese blocks the curds are filled into cheese hoops—perforated boxes in which the curds are formed into a final loaf, block, cylinder, or other desired shape by a cheese press. Pressing also expells additional whey and thus determines the extent of further microbial fermentation in the cheese blocks. Even the shape obtained in pressing is an important determinant for some cheese varieties (e.g. the round balls of Edam).

Curing or Ripening

This is the final and most crucial step in the cheese-making process which will determine the quality of a cheese. While all the preceding operations take about 7 hours, the curing may last for a minimum of a few days to more than a year. Various enzymatic processes caused by the natural milk enzymes, as well as enzymes produced by the microbial cultures, are responsible for the desirable (or sometimes undesirable) flavor developments. The major events occurring during ripening may be the protein breakdown (Camembert), development of gases (holes in Swiss cheese), the breakdown of fat (blue cheese flavor), production of acids and other flavor compounds, etc. The same reaction which is desirable in one type of cheese (e.g. gas production in Swiss cheese) may be a serious defect in other cheeses (Cheddar, Colby).

The ripening usually takes place in ripening rooms where temperature, humidity and other factors must be controlled differently for each type of cheese. Figure 6-6 shows a typical ripening cellar of a small cheese factory in Switzerland. Longer ripening processes give richer flavors, but the final product is more expensive since the long storage costs more.

Processed Cheese

Contrary to a fairly common public view, processed cheese is a valuable dairy food containing much the same nutrients as other cheeses (Table 6-7). The primary reason for developing the processed cheese products has been the need to utilize those natural cheeses which could not be sold through regular markets. Some may have been mechanically damaged during the storage or in the ripening process, others may have developed excessive acidity (or not enough of it) or have other minor defects, making them less attractive—although still perfectly acceptable—for direct consumption. The processed cheese is made by milling the cheese ingredients, mixing with other flavoring or technological ingredients (e.g. emulsifiers to maintain good emulsion stability) and melting the mixture into a homogenous mass by relatively high heat. The "plasticity" of the final product is a result of the heat effect on the milk proteins, as well as the successful emulsification of water and fat contained in the product. The heat effect is not destructive to the milk vitamins, e.g. vitamin A and riboflavin, or calcium and other minerals. The high demand for processed cheese has resulted in some cheese factories specializing in production of mild, unripened (and therefore less expensive) cheese exclusively for process cheese manufacture. In Europe processed cheese is a very popular product, often flavored with many interesting ingredients (bacon, chive, onion, even nuts, raisins etc.), while in Japan almost all domestic or imported cheese production is used for processed cheese manufacture.

6.6 BUTTER

In most countries butter is defined as the product manufactured from dairy cream and containing not less than 80 percent milk fat. The only other permitted primary constituents are water, milk solids, bacterial culture (if cultured cream was used as a source of fat), salt, and an approved food color used to minimize differences in the color of butterfat in summer and winter. In some countries, food legislation specifies

FIGURE 6-6. Ripening cellar for Emmenthaler cheese.

the upper limit for water content of butter as 16 percent—the approximate content resulting from the current technological practices.

Butter-making technology ("churning") is based on rather complex physicochemical phenomena characterizing the inversion of an o/w emulsion (cream) to a w/o emulsion that is butter. The three major stages of the traditional butter-making process shown in Figure 6-7 are combined in modern continuous buttermaking machines that have replaced the old butter churn. A fourth step (salting) is often added during the working of the butterfat. Butter made from cream that has been fermented before churning will be more flavorful due to the presence of the compound diacetyl produced by the culture. In modern butter-making processes, concentrated diacetyl may be incorporated into the

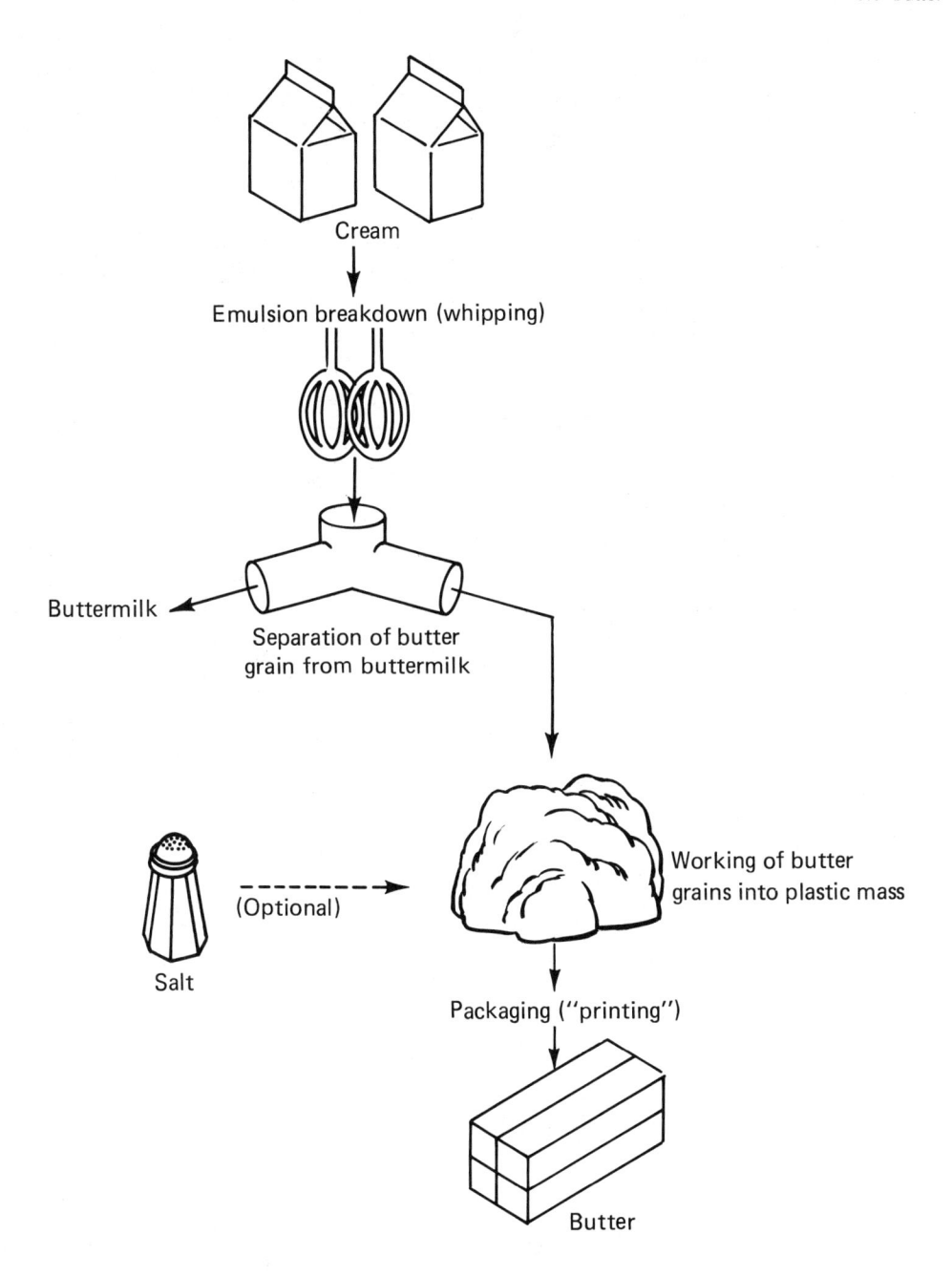

FIGURE 6-7. Schematic illustration of the butter-making process.

fresh cream butter during churning, thus improving the product quality with lower costs than the traditional cream ripening.

The nutritional controversy regarding butterfat, the high and rising prices of butter, and the promotion of polyunsaturated fat margarine, have caused a steady decline of butter consumption in Canada and other countries (Figure 6-8). However, butter is rich in vitamin A, and is one of the least processed table spreads which has been consumed for centuries without any proven ill effects. The allegations linking the normal consumption of saturated fats and particularly butter to increased incidence of coronary heart disease have never been scientifically proven.

The lack of spreadability of butter caused by the high content of saturated fatty acids (Table 5-5), is a serious consumer inconvenience. Much research has gone into minimizing this defect by special manufacturing techniques or by modified cattle feeds resulting in changes in butterfat consistency. In some countries, blends of butter with vegetable oils and/or margarines are legally permitted. The most widely known product of this type is Swedish table-spread "Bregott," consisting of 80 percent butter and 20 percent margarine. The rather conservative North American dairy producers, processors and legislators have been reluctant to initiate changes in the traditionally restrictive

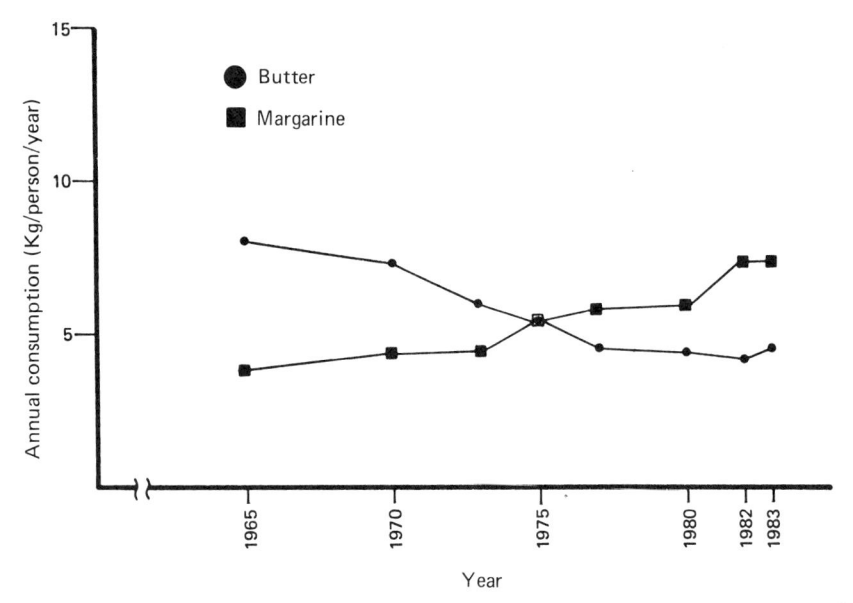

FIGURE 6-8. Per capita consumption of butter and margarine in Canada. (Source: "Facts and figures at a glance," 1984, Dairy Farmers of Canada, Ottawa.)

dairy legislation prohibiting any mixing of butterfat and non-dairy fats in products of the dairy industry, without identifying them as non-dairy imitations.

6.7 ICE CREAM

Although ice cream is eaten as dessert, its significance in human nutrition is unquestionable. The large amounts of ice cream consumed in Canada and the U.S. can be translated into many excess calories resulting primarily from its relatively high fat and sugar contents (Table 6–8), while as a source of protein, vitamins and especially the much needed calcium, ice cream is as valuable as milk.

The principal components of liquid ice cream mix are milk solids and milk fat supplied by fluid and/or powdered milk, cream or butter; sugar and flavoring; and emulsifying/stabilizing materials. Upon freezing, air is incorporated into the mix to give it its light, smooth texture. In North America, frozen ice cream has a typical *overrun* of 100 to 140 percent. Overrun is a measure of air incorporation in the finished product and can be easily determined as

$$OR = \frac{WL - WF}{WF} \times 100\% \qquad \text{(Equation 6–1)}$$

where WL and WF are weights of the same volumes of liquid ice cream mix and finished ice cream, respectively. A 100 percent overrun means that there is about as much air as ice cream mix in a given ice cream container. Various categories of ice cream products (Table 6–8) are differentiated principally by the butterfat content.

Ice cream is a complex mixture of fat globules, the unfrozen solution of milk non-fat solids (Section 10.3), frozen water (ice crystals), and

Table 6–8 Main Types of Ice Cream Products and Their Average Composition

PRODUCT	H_2O	PROTEIN	FAT	CHO[a]	CALCIUM
	----------------------%Total Weight[b]----------------------
Ice cream	62	4	8–12	20	0.15
Ice milk	65	5	3–5	22	0.15
Sherbet	67	1	1	30	0.02

[a]Lactose and sucrose combined.
[b]Average data compiled from various sources.

air bubbles. To achieve and maintain such a fragile multi-component system in optimum balance, the high quality product must be "stabilized" by small amounts of various emulsifiers, thickening agents, vegetable gums, and other food additives. Without proper stabilization the ice cream emulsion can break down during processing and the product may develop various texture defects. Coarseness is a fairly common problem caused by growth of water crystals to sizes detectable in the mouth, while the much more serious sandiness, due to crystallization of lactose in the remaining unfrozen solution, is rare when proper stabilizers are used. Poorly stabilized ice cream would feel very cold in the mouth since much of the water would be frozen. Its melting in the mouth requires large amounts of latent heat supplied by the mouth tissues; such ice cream would also melt rapidly upon standing. Most of these defects can be easily observed in homemade ice creams which can be produced by mixing only the principal ingredients and freezing suitably.

Because of the increasing use of artificial flavors, new emulsifiers and other functional additives, today's ice cream is being occasionally attacked as unnatural and chemicalized. This is obvious nonsense. Ice cream is a purely artifical food resulting from human ingenuity; there are no natural ice cream products. Artifical flavors are used in many other foods today, and their level of usage is normally a fraction of a percent. Much of the flavoring used in ice cream now is man-made. Without artificial strawberry, vanilla or other flavors, however, ice cream would return to the small category of "high-class" foods affordable only by royalty the way it all began several hundred years ago.

6.8 CONCENTRATED, STERILIZED AND DRIED PRODUCTS

Sterilized plain or evaporated milk, sweetened condensed milk, or dried milk products are manufactured to preserve the excess milk produced by farmers. The technologies used are rather complex in engineering design, but simple in principle. Since milk is primarily water, its bulk can be substantially reduced by removing some or all of the water and storing only the milk solids. Furthermore, tying up water with sugar in sweetened condensed milk, or its removal from dried milk powder, will prevent the growth of microorganisms which would be the primary cause of spoilage in these products. Special engineering techniques are used to evaporate the water from the concentrated milk products by boiling at low enough temperatures to minimize undesirable heat effects, or to remove the water from the droplets of milk sprayed into hot air in the spray–drying process. Yet another technique may be used to agglomerate the small particles of dried milk powder into larger grains that will dissolve instantly upon reconstitution with water (Figure 10–3). Chapter 10 gives a more detailed description of these processes.

Another preservation process suitable for extending the shelf life of liquid milk products is heat sterilization. Sterile milk products in sealed glass bottles or cans have been available for many years; their long shelf life is a result of complete destruction of all spoilage microorganisms by heating these products in their sealed containers. However, the "in-container heating" is rather severe. It destroys not only microorganisms, but also greater portions of some micronutrients such as vitamin B_1, B_6, or B_{12}. It also produces a strong, cooked flavor, or even a color change (browning) in the more concentrated products such as evaporated milk. A relatively new industrial technique called "ultra-high-temperature" (UHT) processing overcomes most of these deficiencies by high heat treatment outside the containers, and then filling the shelf-stable milk into germ-free packages aseptically. The heating temperature used is much higher than is possible in the "in-bottle" process. The heating time necessary to kill the microorganisms is thus substantially reduced to no more than 2–4 sec. This highly sophisticated engineering process eliminates the undesirable heat effects on nutrients, and dramatically reduces the development of the cooked flavor and other changes induced in the milk. The final product can be stored without refrigeration for several months. Its eventual decline in quality is usually due to very slow chemical changes resulting in insolubilization of the heated protein. The resulting sediment would be unappealing although perfectly safe. There are no chemical additives used in this process; the long shelf life is achieved only by deactivating all spoilage microorganisms and enzymes by the precisely controlled heating. The UHT process is an outstanding achievement of modern dairy science and technology in combination with innovative package engineering.

6.9 NUTRITIVE VALUE OF MILK AND DAIRY FOODS

Pasteurized milk and other processed dairy products are significant sources of many nutrients, including protein of a very high nutritional quality, several vitamins, and important minerals such as calcium, phosphorus, and magnesium. The estimated contributions that dairy foods make to nutrient supply in North America are shown in Table 6–9.

Pasteurization has a minimal effect on the nutrients in fluid milk. Destruction of small amounts of heat labile vitamins may occur, but this is a necessary trade-off for eliminating the direct health hazard of

Table 6-9 Contribution of Dairy Foods Excluding Butter to Nutrient Levels in United States Food Supply in 1982[a]

NUTRIENT	PERCENTAGE OF TOTAL AVAILABLE FROM ALL SOURCES
Protein	22
Calcium	72
Phosphorus	34
Riboflavin (vitamin B_2)	37
Vitamin A	13
Vitamin C	3

[a]Courtesy of National Dairy Council, Rosemont, Ill. 60018.

the pathogenic microorganisms that may be present. One of the most heat labile vitamins is vitamin C. Milk however, contains very little of this vitamin and is not an important source of it in our diet (Table 6-9). Thus, the partial destruction of vitamin C in milk pasteurization is of no nutritional significance.

Fat-soluble vitamins A and E may be removed from some low-fat milk products by fat separation. In fluid milk sold in North America the vitamin A level is often maintained by voluntary fortification, while vitamin D fortification is required by food laws. The vitamin D fortification is especially important since this vitamin is needed by the human body for proper dietary absorption of calcium and phosphorus, both of which come principally from milk. Also, the only other major source of vitamin D is our own skin when exposed to sunlight, which in many geographical areas is less than plentiful, especially in winter.

Milk in glass bottles or clear plastic "pouches" must be protected from direct sunlight or strong artifical light due to the light sensitivity of riboflavin, another important vitamin contained in milk. Other vitamins, milk proteins, and lactose—a dietary source of energy—are not affected by pasteurization or storage of fluid milk or fermented dairy products.

Processing milk into cheese removes most of the water-soluble vitamins and lactose, and some of the minerals. On the other hand, protein and fat-soluble vitamins are concentrated, and variable amounts (depending on the cheese type) of calcium and phosphorus are also retained by the cheese-making process. In cultured fluid milk products, essentially all nutrients coming from milk are retained, and additional micronutrients may be added by the bacteria during the fermentation process. The alleged magic nutritive values of yogurt, kefir and other fermented milks have not been proven though. The large increases in the vitamin content of yogurt due to microbial fermentation that were claimed in the past by some authors of controversial nutrition literature (e.g. Adelle Davis), appear to be a complete myth unsupported by any scientific evidence.

Fermentation of dairy products also removes a portion of the lactose present in milk. This may make cultured products slightly more digestible to consumers experiencing so-called "lactose-intolerance," a condition characterized by the inability of the body to produce the enzyme *lactase* needed to digest lactose. For those who cannot consume milk at all due to severe lactose-intolerance problems (flatulence, diarrhea), an industrial lactase enzyme preparation is now marketed for home use. When added to milk several hours before consumption, this lactase preparation will split the lactose into its two monosacharide constituents, glucose and galactose, much the same as a body of a lactose-tolerant individual does. The lactose-intolerance condition seems to develop gradually in life, since practically all newborn babies need be highly lactose tolerant (mother's milk contains almost twice as much lactose as cow's milk).

Occasionally consumers complain about general discomfort described as "heavy stomach" resulting from drinking milk but not fermented products. This may be caused by difficulties in digesting casein from fluid milk that in the abrupt change to the acid environment of the stomach may develop a hard clot, as opposed to a soft clot produced by the gradual milk fermentation process. Various dairy drinks based on whey, which in itself used to be a popular drink consumed in special "whey houses" in 19th century Europe, have been gaining popularity recently (Table 6–10). These nutritious whey drinks may offer a dietary alternative to either fluid or fermented milk. They are an adequate source of whey protein, one of the best proteins for human nutrition, and especially calcium, the most significant nutrient supplied by dairy products. Lack of calcium will lead to osteoporosis, a characteristic disease of old age manifested by the weakness of bone. The development of osteoporosis is a gradual process, in which calcium from bone is slowly utilized by a body faced with a chronic dietary calcium deficiency. Thus, the poor dietary habits of youth may have to be paid for much later. Milk and other dairy products are among the best natural sources of easily absorbable calcium, some of the nutritionally most complete proteins and other important nutrients. They should be a regular part of a daily diet for children as well as adults.

6.10 CONTROL QUESTIONS

1. Why is all raw milk tested for antibiotics content before being accepted by a dairy processor? How can the antibiotics get into the milk?

Table 6-10 Characteristics of Commercial[a] Dairy Drinks Based on Whey

PRODUCT	COUNTRY OF ORIGIN	CHARACTERISTICS
Rivella, Surelli	Switzerland	Clear, deproteinated, carbonated, contains 35% whey
Nature's Wonder	Sweden	Mixture of hydrolyzed lactose syrup, whey protein concentrate and special fruit juice blend
Big M	West Germany	Flavored whey enriched with vitamin E
Frusighurt	West Germany	Mixture of whey (90%) and apple juice (10%)
Other	Switzerland, Holland, W. Germany	Mixtures of whey and fruit juices (orange, grapefruit, passion fruit, mango, maracuja)

[a]Market survey by the author, 1983.

2. What is the main purpose of milk pasteurization? What is the function of the regeneration section in the HTST milk pasteurizer?

3. Why is it necessary to pasteurize milk at 90°C for five minutes for yogurt manufacture?

4. How much homogenized milk of typical composition must be mixed with 1,000 kg of skim milk to produce "partially skimmed" (2% butterfat) milk? How much final product will be made if we have 50,000 kg of homogenized milk and unlimited supply of skim milk available for this product?

5. What is the difference between curd production in cottage cheese and cheddar cheese manufacture? pH.

6. Imagine that you were just hired by the "Jolly Moo" Dairy Company as a quality control manager. On your first day a cheese room operator rushes into your quality control lab and asks you for help. He apparently has a problem with setting cottage cheese - even after 6 hours the milk in the 2,000 gal vat did not coagulate (did not form the gel). He claims that he followed exactly the procedure used for cottage cheese making but he is very inexperienced. List all possible reasons (including human-

error-type ones) that might have caused this failure. What quick tests would you do to help your unfortunate operator salvage the 2,000 gal of milk?

7. Explain the reason for fortification of fluid milk with vitamin D.

Suggested reference books—
Milk and Dairy Products

Alfa-Laval, A. B. *Dairy handbook.* Lund, Sweden, undated.

Arbuckle, W. S. 1977. *Ice-cream.* Westport, Conn.: AVI Publ. Co.

Harper, W. J. and Hall, C. W. 1976. *Dairy Technology and Engineering.* Westport, Conn.: AVI Publ. Co.

Kessler, H. G. 1981. *Food Engineering and Dairy Technology.* Freising, W. Germany: Verlag A. Kessler.

Kosikowski, F. V. 1977. *Cheese and Fermented Milk Foods.* Ann Arbor, Mich.: Edwards Bros.

Lampert, L. M. 1970. *Modern Dairy Products.* New York: Chemical Publ. Co.

Paige, D. M. and Bayless, T. M. 1981. *Lactose Digestion.* Baltimore, Md.: The Johns Hopkins University Press.

Quebec Dairy Foundation. 1984. *Dairy Science and Technology.* Ste. Foy, Que.: Laval University Press.

Renner, E. 1983. *Milk and Dairy Product in Human Nutrition.* Munich, W. Germany: Volkswirtschaftlicher Verlag.

Walstra, P. and Jenness, R. 1984. *Dairy Chemistry and Physics.* Toronto: J. Wiley & Sons.

Webb, B. H., Johnson, A. H., and Alford, J. A. 1974. *Fundamentals of Dairy Chemistry.* Westport, Conn.: AVI Publ. Co.

Webb, B. H. and Whittier, E. O. 1970. *Byproducts From Milk.* Westport, Conn.: AVI Publ. Co.

f.

7

Processing of Meat, Poultry and Fish

7.1 SOURCES AND PRODUCTION

In the broadest sense, the sources of foods belonging to this category include muscles, edible organs and other parts of certain agricultural animals as well as poultry, fish, wildlife and other living creatures. However, processing of *meat*—meaning traditionally beef, pork and mutton or generally *red meat*—is usually considered separately from poultry or fish processing, since the corresponding technologies have developed separately. In this text the common biochemical and technological principles of processing animal muscles and organs into human food will be explained primarily for the red meat industry.

Red meat production is one of the least efficient agricultural processes in terms of conversion of raw material to human food (Table 7-1). In many instances, however, cattle grazing is the only productive use of noncultivatable agricultural land. Ruminants (beef cattle and sheep) are especially useful for their capability of utilizing cellulosic forages that can be grown on marginal lands. In a controlled feeding operation (feedlots), the conversion of cattle feed to human food can be more efficient and the time needed for optimal carcass development reduced, resulting in better returns for the producer and savings for the consumer. Feeding of fully developed animals is wasteful since little increase in lean muscle mass will occur beyond a certain stage of maturity (Figure 7-1). This is similar to the required amounts and utilization of food consumed by a hungry growing child, an active adolescent or a mature sedentary adult. Approximate requirements for conversion of

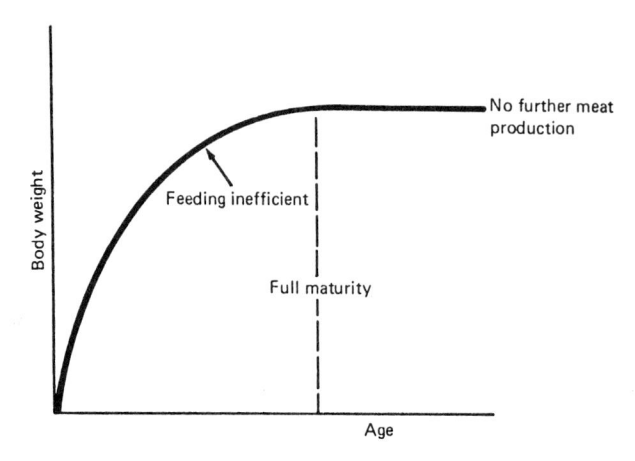

FIGURE 7-1. Relationship between weight gain and maturity of meat-producing animals.

animal feed to human food for beef, pork, and poultry production are included in Table 7-1.

7.2 MEAT COMPOSITION

Meat (meaning red meat in this text unless otherwise indicated) is usually defined as the edible part of the skeletal muscle of an animal that was healthy at the time of slaughter. Other edible parts of an animal slaughtered for food (e.g. livers, kidneys, blood, etc.) are referred to as *meat by-products*.

Table 7-1 Land and Feed Requirements for Production of Edible Food Protein

	PRODUCTION REQUIREMENTS[a]	
	kg FEED/kg MEAT	m^2/kg PROTEIN/Year
Meat Source		
Beef	7–10	300–500
Pork	4–6	200–400
Chicken	2–3	250–300
Plant Protein Sources		
Soybeans	–	30
Potatoes	–	40
Cereals	–	80

[a]Estimates by author from various sources

The principal three components of red or poultry meat are water, protein and fat (Table 1–3). Their proportions vary widely according to the species of animal, its age, nutrition status, and the location of the meat on the carcass ("cut"). The amount of moisture contained in the meat is largely a function of the water-holding capacity of meat protein. An approximate relationship between moisture (M) and protein (P) contents of a given meat cut is

$$P = \frac{M}{4} \qquad \text{(Equation 7–1)}$$

Major types of meat proteins and their principal constituent groups are given in Table 7–2.

The mineral content of deboned meat is very low, and the carbohydrate content of all meats is negligible. The principal animal carbohydrate *glycogen*, which is normally present in muscles of living animals in about 1 to 2 percent concentrations, is used up shortly after slaughter in various complex biochemical reactions occurring in the transitory state of *rigor mortis*.

Table 7–2 Principal Protein Constituents of Deboned Meat

COMPONENT GROUP (% OF TOTAL MEAT PROTEIN)	MAIN COMPONENTS	PROPORTION OF COMPONENTS[a] WITHIN A COMPONENT GROUP (%)
Muscle (myofibrillar) proteins (60%)	Myosin	56
	Actin	22
	Tropomyosin	13
Plasma (sarcoplasmic) proteins (30%)	Glycolytic and other enzymes	80
	Myoglobin and residual hemoglobin	15
Connective tissue proteins (10%)	Collagen	50
	Other tissue proteins	50

[a]Estimates by author from various sources.

7.3 LIVING MUSCLE AND RIGOR MORTIS

Conversion of muscles of a living animal to meat or a meat product can be viewed as a simple mechanical "disintegration" of the carcass after the animal is slaughtered. Living muscle, however, is an extremely complex biochemical phenomenon. The numerous biochemical reactions governing the actions of the muscle during life do not stop immediately after the animal is killed. At least cursory knowledge of the most significant reactions occurring after the slaughter is necessary to understand some of the basic steps in fresh and processed meat technology.

In the living muscle, one of the many important biochemical reactions involving generation of energy needed for proper muscle function is generally termed *glycolysis* (Equation 7-2). The animal "equivalent of starch," a carbohydrate called *glycogen*, is used up to generate the needed energy in the presence of O_2:

$$\text{glycogen} \xrightarrow{\text{enzymes}} \text{glucose} + O_2 \xrightarrow{\text{enzymes}} CO_2 + H_2O + \text{energy}$$

(Equation 7-2)

When O_2 is not available (after the animal is killed or under stress) the glycogen and the resulting glucose are broken down to *lactic acid* which has a substantial effect on the pH drop in the muscle. In a living organism, the lactic acid can be eventually utilized along another biochemical pathway, thus keeping the living muscle above pH 6.0. After death this alternative pathway will not be utilized, and the pH of the meat may drop to about 5.4 or lower. The increased acidity gives the fresh meat some protection against microbial spoilage. As a result of this and many other changes, a set of biochemical reactions typical of the "dying muscle" will ensue some time after slaughter, depending on the animal, its size, and other factors. The muscle is said to undergo a state of rigor mortis. Of all the various effects of rigor mortis, the most important one for the meat processor is the stiffening of the muscle fibers resulting in a temporary but substantial toughening of the meat. If the meat is cooked or frozen during rigor mortis, the eating quality of the final product will be poor since the toughness cannot be easily overcome by conventional cooking procedures.

After completion of rigor mortis (several hours for chickens and up to a few days for the large "red meat" animals), the meat slowly returns to—and often exceeds—its pre-rigor tenderness. This is caused by various proteolytic enzymes present in the meat. In the absence of controlling biochemical mechanisms of the living organism, the endogenous enzymes slowly degrade the muscle fibrous structure. In controlled industrial conditions we speak of *aging*, which is an important step in fresh meat technology and is used particularly in handling beef.

The time of aging is largely dependent upon temperature as shown in Figure 7-2. From an economic standpoint, an important feature of the aging step is the protection against water loss, since every pound of water evaporated means a loss of revenue for a pound of meat. For this

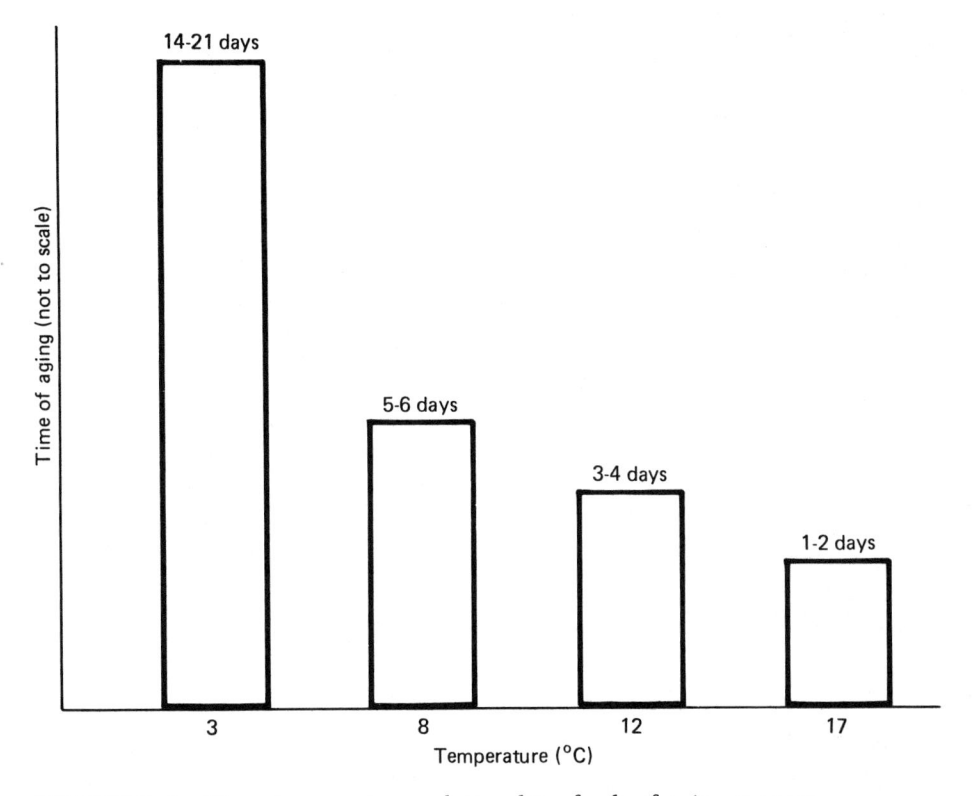

FIGURE 7-2. Time–temperature relationships for beef aging process.

reason, the aging is usually carried out at a low temperature and about 100% relative humidity under strict microbiological control. Various improvements of the relatively slow and costly aging process have been suggested by researchers. Electrical stimulation of the freshly slaughtered carcasses by low voltage electricity has been recommended as a pretreatment for improved tenderization results. Although still the subject of ongoing research, this treatment appears to be gaining industrial acceptance as it results in improved eating quality of certain meat cuts for the consumer.

The function of meat tenderizers, sometimes used at home and in the restaurant industry (but not in retail of fresh meat), is based on the same principle of enzymatic breakdown as in natural aging. The tenderizing enzyme preparations, however, are more powerful and often less discriminatory of the kind of meat protein (Table 7-2) they break down. The action of an improperly used tenderizer can cause a "mushy," unnatural texture in cooked meat.

7.4 FRESH MEAT

The technology of fresh meat production is simple. After slaughter (usually by electric stunning followed by slitting the throat and hanging by the hind legs to drain all blood) the carcass is skinned, opened by a mechanical saw, and eviscerated (Figure 7-3). Government health inspectors approve every carcass as suitable for human consumption at this point by visual inspection of the viscera and certain other parts which would show any characteristic health-related malformations. The approved carcasses are then moved to aging rooms, and after aging broken down to large blocks with a chain saw. These primal cuts are further disintegrated to the individual retail cuts by hand labor. Fresh meat processing may be compared to a car assembly line in reverse, including the high density of workers (Figure 7-3).

Alternatively, the disintegration of the carcass may be carried out shortly after the slaughter when the meat is still at the animal body temperature ("hot", hence "*hot deboning*"). The hot deboning is now commonly used by many meat processors. The advantage of pre-rigor deboning is the space and energy savings achieved by not having to refrigerate and transport the bones as a part of the whole carcass. The

FIGURE 7-3. Carcass breakdown and evisceration.

rigor mortis and subsequent aging then take place in meat already packaged for transportation in boxes ("*box-ready*" meat).

The labor-intensive hand-deboning is costly and not highly efficient. About 7–8 kg of meat is normally left on the hand-deboned carcass due to the difficulties in removing portions of meat clinging to the bone. A relatively new mechanical separation ("deboning") technology (Figure 7–4) can be used to recover at least some of this waste meat by grinding the bones into a mixture of meat protein paste and bone particles. The meat paste is separated from the bone particles by squeezing it through a fine sieve or other mechanical separating device mounted at the end of powerful grinders. The microscopic particles of bone that may be included with the protein paste are harmless to consume and could be a very useful source of dietary calcium, one of the most important minerals in our diets. The mechanically separated protein paste is suitable as an ingredient for sausages, luncheon meats, and other processed meat products. Using too much of the mechanically separated meat may impart a slightly gritty texture to some of the products.

FIGURE 7–4. Mechanical deboning machinery for separation of meat protein paste from crushed bone. (Courtesy of Protein Foods Corp., Hamilton, Ontario.)

The main parameters of fresh meat quality used by the consumer are meat *color* and fat *marbling*, indicating the total amount and distribution of the fat. These two factors are also used by government agencies for grading of meat quality—a somewhat superficial and often confusing process. The grading of beef is based on visual observation of the meat and the surrounding fat exposed by a cut between two adjoining ribs. The color and the texture of the exposed cross-section of the *longissimus* muscle ("rib-eye"), the appearance and color of the total fat layer on the outside of the carcass and certain visual indicators of maturity will determine the appropriate grade (Table 7-3). The numerical part of the grade (i.e. A1-A4) will be assigned according to the thickness of the fat surrounding the rib-eye. Figure 7-5 illustrates the measurements used for beef grading. A similar visual procedure is used for grading other meats, as well as poultry. The "lower" grade (i.e. B grade of beef) may be just as flavorful and nutritious as the A grade—only it does not look as good!

The color of red meats is determined by the chemical state of the purplish red muscle pigment *myoglobin* (as opposed to the blood pigment *hemoglobin*). The myoglobin, whose main role in living muscle is to bind oxygen needed for the various chemical reactions, can exist in two states. In the presence of oxygen, the myoglobin is oxygenated to *oxymyoglobin* and the muscle is bright red in color. When much of the oxygen is removed, the myoglobin will change to *metmyoglobin* with the resulting brownish color developing in the muscle. The creation of metmyoglobin by vacuum packaging in improper semi-permeable plastic packaging materials, although harmless, may be a serious merchandising problem since consumers associate the brownish meat color with old products. Red meat cuts sold at supermarket stores are usually wrapped in oxygen-permeable plastic foil to avoid this problem.

Cooking meat denatures the meat pigment, as well as the various meat enzymes involved in its color changes in the fresh state. This gives cooked meat its characteristic greyish-brown color. Cooking procedures will also accomplish tenderization of meat obtained by heat softening of the connective-tissue protein, collagen. A temporary toughening effect caused by heat coagulation of some of the muscle proteins may precede the collagen tenderization during certain cooking methods. A relatively short period of desirable tenderization may be observed at the very beginning of the heating process if sufficiently high temperatures are used, as in preparation of minute steaks. Other effects of meat cooking include the *shrink*, i.e. loss of juice due to the loss of *water-holding capacity* of the heat-denatured muscle proteins, the development of desirable flavors (the complicated chemistry involved is still poorly understood), and safety aspects. Particularly important in this regard is the need for destruction by heat of path-

Table 7-3 Quality Indicators Used for Beef Carcass Grading[1]

Grade	Maturity	RIB-EYE MUSCLE Color	RIB-EYE MUSCLE Texture	RIB-EYE MUSCLE Marbling	CARCASS COVER FAT Color	CARCASS COVER FAT Texture	CARCASS COVER FAT Distribution
A	Youthful	Bright red	Firm, fine-grained	At least slight	White or reddish hue	Firm	Uniform, or slightly lacking on hip *or* chuck
B	Youthful	Bright red to medium dark	Firm or slightly coarse	No minimum	White to yellowish	Firm to slightly soft	Somewhat lacking on hip *and* chuck
C	Youthful to intermediate	Bright red to dark	Fine-grained to coarse	—	White to pale yellow	Firm to soft	At least slight cover
D	Mature	—	—	—	White to deep yellow	Firm to soft	Slight to excessive

[1]Source: Agriculture Canada specifications.

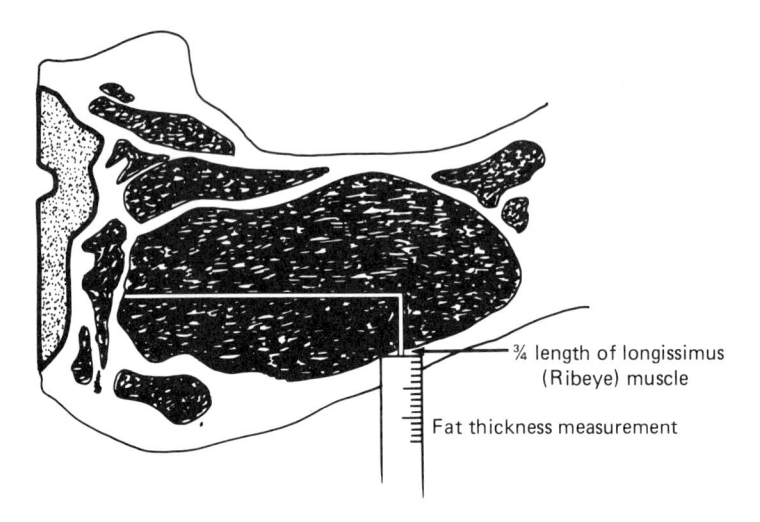

FIGURE 7-5. Measurement of fat thickness for beef grading.

ogenic bacteria of the *Salmonella spp.* associated commonly with chicken. Proper heating also destroys larvae of the muscle-invading parasite *Trichinella spiralis*, traditionally considered a potential although rare hazard with pork (one reason for barbecuing pork to the well-done state!) but often found in muscles of some large game animals such as bears.

Individual cuts of meat will produce different eating qualities and will respond differently to various heating conditions since the structure of individual muscles differs with their intended biochemical function. Figure 7-6 shows the position of major cuts on the beef carcass. The large differences in price of retail cuts reflect mainly the "prestige" factor, as well as the flavor and texture difference after cooking, and the differences in proximate composition (Table 7-4). Often the cheaper cuts could have higher nutritive and economic values due to their protein content, lower shrink and other aspects.

7.5 CURING

One of the first processing techniques used at home and by industry to preserve an oversupply of meat was *curing*. In addition to the preservation effect, curing alters the flavor and color of treated meat, and expands the selection of meat products available to the consumer.

The principle of preservation by curing is the infusion of the meat with certain salts and sometimes other functional ingredients used for improvements of flavor, reduction of water loss and other quality

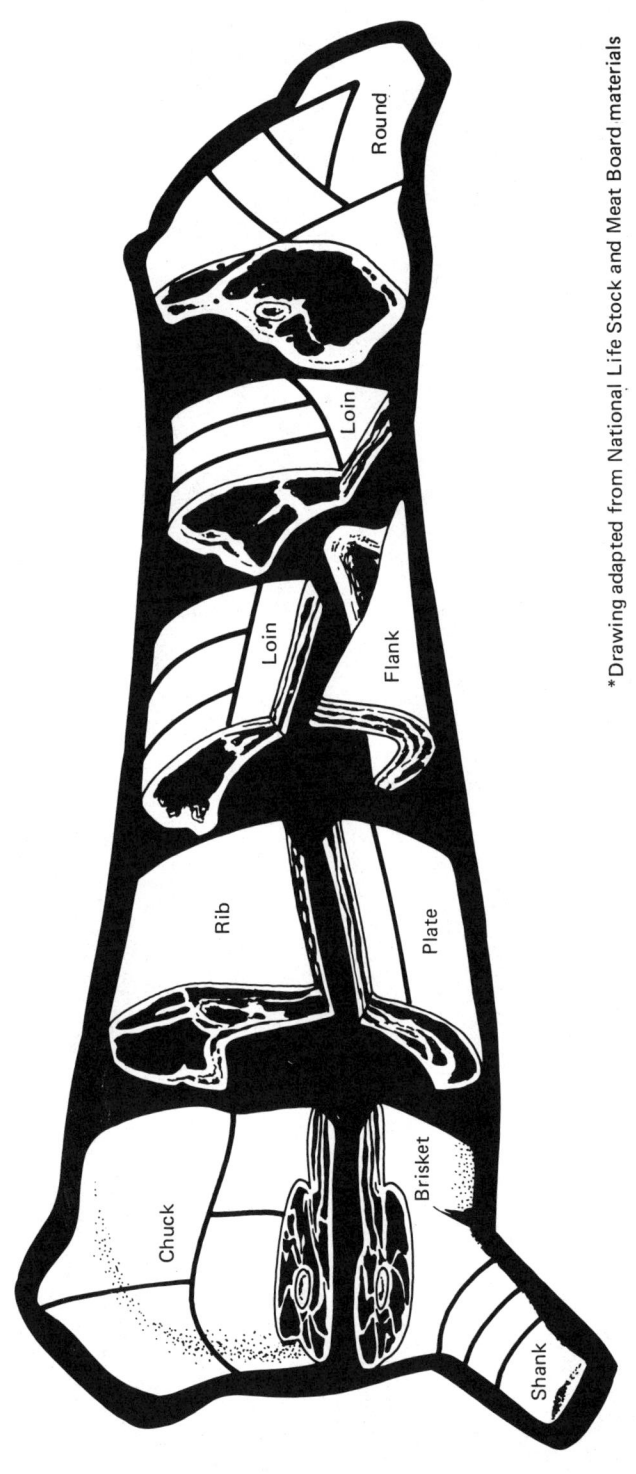

FIGURE 7-6. Position of various cuts on the beef carcass.

*Drawing adapted from National Life Stock and Meat Board materials

Table 7-4 Proximate Composition, Bone Content and Retail Value of Selected Beef Cuts[a]

Retail Cut	COMPOSITION (%) H₂O	Protein	Fat	PERCENTAGE OF TOTAL WEIGHT Edible Meat[b]	Bone	RELATIVE PRICE[c] Edible Meat	Protein
Rib roast	53.3	13.5	25.0	74.7	7.4	1.6	1.4
Brisket (deboned)	57.7	15.2	26.0	83.0	0.2	1.5	1.4
Ground chuck	61.9	16.7	20.3	100	0.0	1.0	1.0
T-bone steak	44.7	12.4	28.6	65.6	13.6	2.3	2.1
Boneless rump roast	61.2	16.3	20.9	85.5	0.6	1.8	1.6
Sirloin tip steak	65.6	17.5	15.8	90.2	0.0	2.0	1.7

[a]Source: Funk et al., J. of CIFST 9(1976):35. By permission of Canadian Inst. Food Science Technol., Ottawa, Ont.
[b]After removing trimmable fat, bone and inedible connective tissue.
[c]Basis-prices for ground chuck. Comparisons of equal weights.

aspects important to consumers. In the early days, salt may have been added by rubbing the meat pieces with salt crystals ("corns", hence "corned beef"). In current industrial practice, the salts are added as a water solution ("pickle"), containing several complementary curing ingredients that may be used. The meat may be soaked in the pickle in curing vats, or the pickle may be pumped into the piece of meat through one of the arteries (picnic hams), or by a grid of needles (bacon).

The characteristic pink color of cured meats, wieners, sausages and many other meat products processed with the cure comes from the principal ingredient *sodium nitrite* (NaNO$_2$). The main visual curing effect of this salt is its color fixation capability according to the following *curing* reaction presented here in simplified principle only:

$$\text{myoglobin} + \text{NaNO}_2 \xrightarrow{\text{reducing agents}} \underset{\text{myoglobin}}{\text{nitric oxide}} \xrightarrow{\text{heat}} \underset{\text{chrome}}{\text{nitroso-hemo-}}$$

(purplish red)   (bright red-unstable)   (pink-stable)

(Equation 7–3)

The heat-stable pigment nitroso-hemo-chrome is produced from the nitric oxide myoglobin by heat (as in a smoke house) and gives the cured meat its typical color which will not fade upon cooking.

The sodium nitrite is included in the curing reaction as the source of nitric oxide (NO) which in turn is needed for the color change. The sodium nitrite is not completely used up in the curing process. The residual amount of nitrite remaining in the meat after the curing is strictly regulated since nitrite has an important preservative function. It is the only food additive known to be effective against the deadly microorganism *Clostridium botulinum* associated with foods and particularly meats.

The use of NaNO$_2$ by the meat industry has become one of the most controversial issues in all of industrial food processing. In some countries (e.g. Norway) the use of NaNO$_2$ in meats is now prohibited. It has been shown that upon severe heating, nitrite may combine with certain proteins (such as meat protein) to produce *nitrosamines* known to be carcinogenic in experimental animals. A health hazard due to the presence of nitrosamines in cured meats which may be severely heated (such as bacon), however, has not been scientifically proven. Furthermore, there is research evidence that, in our guts, our body may produce nitrosamines from the protein content of our food with no nitrite ingested whatsoever. The amount of nitrite present in the human body (saliva), in some agricultural food products (spinach), or in the polluted air that we breathe in big cities, may be several times higher than the residual nitrite permitted in the cured meats as the preservative against *botulism*. The precise protective mechanism of $NaNO_2$ against *C. botulinum* growth is not well understood, and a suitable alternative material has not been found so far. Thus, the $NaNO_2$ controversy is a typical example of the risk-benefit situation with which the food regulatory agencies are faced in their decision-making process. Banning $NaNO_2$ in meat processing would reduce the selection of meat products available to the consumer and would ultimately decrease meat production with corresponding price increases. The utilization of lower grade meats and meat cuts would become more difficult. Elimination of the curing process, which, in essence, is the basis for the processed meat industry, could eliminate thousands of jobs and millions of dollars in corporate taxes redistributed to us all in various government-funded programs. The continued use of $NaNO_2$ may not be detrimental to human health in general as evidenced by the long history of nitrite presence in processed meat products. Should it be allowed, however, if there is even the smallest risk of possible carcinogenicity for someone who may be extremely susceptible?

The current solution of government regulatory agencies is to limit the maximum allowable input of $NaNO_2$ into the cured meat products. In Canada, not more than 200 ppm (150 ppm in the case of bacon) is allowed to be added prior to the curing process. This greatly reduces the danger (real or imaginary) of excess residual nitrite in the final products which could conceivably result in formation of the nitrosamines when the product is heated to a high temperature by the consumer. Yet there is still enough antibotulinal activity provided by the remaining $NaNO_2$ for product safety.

7.6 PROCESSED MEAT PRODUCTS

Endless varieties of sausages, salami, blood puddings, head cheese, liver paste, and other similar products are manufactured by the meat industry worldwide. The main ingredients used in these comminuted products are meat trimmings and cuts not marketed as fresh meat, lower grade carcasses (see Table 7–3), certain specified meat by-products such as tripe, liver, blood or blood plasma, and non-meat ingredients called *fillers* or *binders*. Some of the most typical fillers are listed in Table 7–5; the mixture of a filler with a sugar (sucrose, glucose, corn syrup solids), salt, spices, egg white or other gelling agents, and a preservative such as sodium nitrite, is defined as binder.

The food laws of some countries (Canada and the U.S.) are strict about the maximum amount of filler or binder permitted in finished prepared meat products; in other countries (e.g. South Africa) there is no limit on the use of non-meat ingredients. Legal limits in terms of

Table 7-5 List of Fillers and Binder Components Used in Processed Meat Products

FILLERS	ADDITIONAL COMPONENTS FOR BINDERS[a]
Grain or potato meal	Salt
Processed wheat flour	Sugar (dextrose, sucrose, etc.)
Bakery products (bread, biscuits)	Corn syrup solids
Milk or whey powders	Egg albumin
Starch	Curing agent (if permitted)

[a] A binder is a mixture of any filler plus one or more of the additional components listed.

maximum amount of water and fat and minimum amount of protein content in the final product are also established for various products. Examples of legal requirements for some typical meat products are listed in Table 7-6.

In the United States, a general rule for maximum allowable moisture content (% by weight) of processed meat products is

$$M = 4P + 10 \,(\%) \qquad \text{(Equation 7-4)}$$

Thus, the amount of moisture (M) in a cooked salami containing 14% protein (P) must not exceed 66% in the USA. In Canada, moisture content in products containing fillers is limited to 60%. The maximum fat content of cooked sausages is limited to 30% in both countries.

The choice of principal meat ingredients is a matter of the recipe, the technological (or *functional*) properties of the individual meats, and economy. Numerous calculations are performed by the meat pro-

Table 7-6 Examples of Standards of Identity for Classes of Prepared Meat Products[a]

PRODUCT[b]	% PROTEIN (MIN)	% FAT (MAX)
Fresh sausage (uncooked)	9	40
Products containing filler and pork	13	40
Any other product containing filler	13	30
Products containing seasoning and pork	16	40
Any other product containing seasoning	16	30
Meat loaf	11	Unspecified

[a] Source: Food and Drug Act and Regulations, Ottawa.
[b] According to Canadian food laws, no prepared meat products with filler may contain more than the dextrose equivalent of 4% of the filler nor more than 60% moisture.

cessor almost daily to adjust any particular formula to the rapidly changing market situation regarding prices of meat. These so-called *least-cost* formulations take the form of a series of linear equations solved by routine computer programs. Several dozens of components will generally be available for any particular formula. The needed amounts of the few components finally selected will be determined from the composition of the individual components available, the composition and quality requirements for the final product, various functional properties (e.g. water binding, fat binding, color, heat coagulability) of individual ingredients, and the cost of the final formula. Table 7-7 lists some frequently used ingredients and their properties used in least-cost formulations.

The technology of prepared meat products is varied but typically includes *grinding* ("chopping"), *mixing* (formulation), *stuffing* the final pasty mix into natural (washed and sanitized guts) or artificial (collagen, cellulose or plastic film) casings, and cooking in hot water bath or spray or in a smokehouse. The cooking step coagulates the meat protein, thereby setting the final structure of the product. It also com-

Table 7-7 Example of Data Needed in a Least-Cost Formulation Process for Processed Meats

A. REQUIREMENTS FOR A FINISHED PRODUCT

Total Weight	100 kg
Maximum fat content	30%
Maximum moisture content	60%
Minimum protein content	13%
Maximum collagen content	35%
Minimum beef content	30%
Minimum hearts content	10%
Amount of spice used	6%
Maximum added water	10%

B. AVAILABLE INGREDIENTS AND THEIR CHARACTERISTICS

	Relative Cost	Fat	Moisture	Protein	Collagen in Protein
			---------- % ----------		
Pork trimmings	.4	55.0	34.4	9.9	34.0
Beef shank meat	.9	7.1	72.7	18.9	66.0
Beef head meat	.6	13.5	67.9	17.2	73.0
Pork hearts	.5	15.0	68.9	15.7	27.0
Spice	.5	–	10.0	–	–
Water	–	–	100.0	–	–

pletes the curing reaction by heat-stabilizing the nitroso-hemo-chrome pigment (Equation 7-3), and provides the desirable flavor of smoked meats. For products cooked in hot water (as in the manufacture of wieners), the aroma compounds extracted from smoke of a burning wood may be added as a "liquid smoke." Hot smoke, generated by burning moist sawdust or hardwood, is used in modern smokehouses with strict temperature and humidity control. The heat-processed products may be skinned (as in the case of wieners where the plastic casing is removed after the cooking step), sliced, vacuum packaged or sold as such.

A great deal of scientific knowledge as well as practical empirical observations are used in the technology of prepared meat products. For example, certain kinds of proteins from some parts of some animals are much more useful for the emulsification of the fat in the product than other parts or other animal proteins. Technologically improper ingredients and/or processes like overchopping or excessive smokehouse humidity may result in serious quality defects. Formation of undesirable fat caps, or worse yet, gelatinous pockets, may result from the breakdown of meat emulsion or the connective tissue protein collagen in the heating process. Proper use of the various fillers and binders is also very important in this respect, as is the proper choice of various types of meat (beef, pork, mutton) for final flavor and emulsion stability.

7.7 POULTRY PROCESSING

In the preceding paragraphs the basic aspects of meat processing were described predominantly for red meats. Some of the biochemical considerations are generally applicable also to poultry. Contemporary chicken and turkey meat production is one of the most efficient operations in converting agricultural materials to meat (Table 7-1). As a result, chicken meat is among the most economical buys for consumers, especially when compared on the basis of the cost of protein.

The compositional similarity of poultry and red meats as to the three predominant components was shown in Table 1-3. Poultry meat (especially chicken) is significantly lower in fat than most red meats. A major visual difference between the two meat types is the lack of red coloration of fresh—and the whiteness of cooked—poultry meat, resulting from much lower myoglobin content.

Little processing of chicken, turkey or other poultry materials into prepared products is done in North America. The main technological operations are cleaning, eviscerating and packaging for fresh market, or further processing by one of the typical preservation technologies (canning or freezing) which are discussed in Chapter 10. The increasingly common marketing of cut-up poultry results in a surplus of backs and necks, which are often processed by the mechanical deboning machinery. Since poultry bones are much softer than red meat bones, the mechanical separation process is much easier, and may include whole lower grade birds such as spent layers. The poultry protein paste is used as an ingredient in various processed meat products.

A major health concern in poultry processing is the well-known association of pathogenic microorganisms of the Salmonella group with the poultry production environment. Extreme care must be taken in industrial processing or home preparation of poultry not to cross-contaminate ready-to-eat, cooked products with raw materials. A frequent cause of Salmonella epidemics may result from using a knife or cutting board at home for evisceration of raw turkey and later, without careful sanitation, for portioning the cooked bird for Thanksgiving dinner. Salmonella is not a heat-resistant organism; properly handled and/or cooked poultry constitutes no health hazard, but serves as a delicious source of high quality nutrients.

7.8 FISH AND SEAFOOD

Contemporary processing of fresh water and marine fish is determined by the uniqueness of this food source, its availability, and fishing technology, which has been developing rapidly. Huge trawlers with enormous fishing nets are common in marine fishery and "floating factories" often follow the fishing fleets for the immediate chilling or freezing of the catch for further on-shore processing or for direct marketing. The ever increasing popularity of fish foods in the world has been the main catalyst for keen competition and sometimes inconsiderate overfishing of previously rich fishing grounds located in international waters. Most countries, including Canada and the U.S., have instituted a 200-mile territorial rights area around their shores to better control the amount of fish available and the fishing activities. Modern fishing techniques often result in relatively large amounts of *by-catch* ("trash fish") since only relatively few fish species are sought as desirable human food. Finding new ways for utilization of fish by-catch is one of the many tasks awaiting researchers concerned with advances of this relatively traditional segment of the food processing industry.

One of the characteristic problems associated with processing fish and other forms of aquatic life used as seafood (Table 7–8) is the high potential for spoilage. The perishability of fish is notorious, due mainly to the psychrophilic microflora associated with the natural fish environment. A rapid chilling to temperatures just above freezing is necessary to retard the growth of these microorganisms. Some enzymatic

Table 7-8 Principal Types of Aquatic Animals Used As Seafood

GENERIC GROUP AND EXAMPLES	POSSIBLE USES	TYPICAL PRODUCING COUNTRY
Fish - General Use:		
Tuna, cod, halibut, salmon, many others	Fresh, frozen, canned, further processed	Norway, Japan, Canada, U.S.A.
Fish - Special Uses:		
Sardines	Canning	Canada, Portugal
Anchovies	Fish protein concentrate	Peru
Fish eggs	Caviar	U.S.S.R.
Fish by-catch	Kamaboko (fish paste)	Japan
Crustaceans:		
Shrimp, lobsters, scallops, prawns, crabs	Fresh, canned, frozen	Canada, South Africa, Japan, U.S.A.
Other Aquatic Life:		
Shark	Fish and chips	New Zealand
Whale	Fresh, canned	U.S.S.R.
Squid, octopus	Fresh, marinated	Japan

reactions causing protein or fat decomposition, however, may be stimulated at temperatures around the freezing point. These chemical spoilage reactions are caused by enzymes present in the tissue or the digestive tract of the fish and are another reason for the perishability of seafood. Yet other possible causes are the proneness of the highly unsaturated fish fat to rapid oxidation and the exhaustion of glycogen in the muscle from the struggle of the catch.

Further processing of fish in North America is much less sophisticated than the various red-meat industry processes, and is limited mainly to freezing fresh or breaded and par-fried products, as well as canning of specific fish (salmon, tuna, sardines). In some countries, especially Japan and Norway, an extensive and diversified market has been developed for sausage-like fish products, spreads, puddings, marinated or restructured products (such as surimi or kamaboko) and other creations of the aggressive fish-processing industry. In Canada, the leading fishing nation in the world in terms of value of net fish exports, the processing and consumption of fish is rather insignificant compared to other countries with highly diversified food and fish processing industries (Table 7-9). Fishing and aquaculture (fish and seafood farming) should be considered among the most promising sources of food for the ever-increasing population of our planet.

Table 7–9 Total Fish Catch, Total Exports, and Consumption of Fish For Major Fishing Nations[a]

COUNTRY	TOTAL CATCH (1,000 TONNES)	TOTAL EXPORTS (MILL. $ U.S.)	PER CAPITA CONSUMPTION (kg EDIBLE WEIGHT)
Japan	10,400	900	36.4
Portugal	270	122	22.8
Sweden	240	<100	20.8
Norway	2,400	975	18.0
Spain	1,200	340	17.0
United Kingdom	800	360	12.8
France	760	320	12.0
Canada	1,400	1,070	7.0
United States	3,600	990	5.9

[a]Source: Annual Statistical Review, Fisheries Canada, Ottawa. Courtesy of Dr. G. Bligh, Tech. Univ. of Nova Scotia, Halifax.

7.9 NUTRITIVE VALUE OF RED MEATS, POULTRY AND FISH

The main significance of all meats as a human food is their protein content. Being of animal origin and thus close to the "building blocks" of our own body, the meat protein provides all essential amino acids that we cannot synthesize ourselves. Consumers who exclude meat products from their diet need other sources of animal protein (dairy products, eggs), or a carefully selected balance of plant protein sources. Plant proteins are usually deficient in one or more of the essential amino acids.

A second major macronutrient provided by meats is fat. Its main significance is as a rich source of energy (Table 1–7) and thus its intake should be controlled. Some types of animal fat may contain relatively high proportions of saturated fatty acids and less polyunsaturated acids than vegetable fats. In terms of total saturation, beef fat is more saturated than pork, while poultry, and especially fish fats, may be highly unsaturated.

From the micronutrient standpoint, red meat is a very important source of iron. In the well-fed population of North America, iron deficiency has been identified as one of the very few possible causes for concern especially for certain high-risk groups (pregnant women and adolescent girls). Certain meat by-products (blood, liver) are particularly good sources of easily absorbable dietary iron. Poultry and fish meat contain less myoglobin and thus lower amounts of iron than the red meats; instead, fish is an excellent source of iodine. Red meats and poultry are rich in some water-soluble vitamins of the B complex, such as thiamin (especially abundant in pork), riboflavin and niacin. The main importance of fish as a source of vitamins is the high content of vitamins A and D in the fish liver oil, the dietary supplement dreaded by children before the era of milk fortification (Section 6.9).

The high consumption of foods of animal origin is a typical phenomenon of affluence. The developments in the meat processing industry world-wide have kept pace with the increasing demand for this excellent and nutritious food. The recent technological advances in mechanical deboning, meat product formulation, development of new poultry products, fishing technology and processing, and other aspects of conversion of animal flesh to human food have increased the availability of these products to consumers in all economic groups.

7.10 CONTROL QUESTIONS

1. What is "rigor mortis" and why is it of importance to the meat industry? Is it of the same importance to the fish processing industry? Why or why not?

2. Give a realistic estimate of the proximate composition of a raw boneless sirloin steak weighing 450 g and containing 60 g protein.

3. Are the terms "hot deboning" and "mechanical deboning" synonymous? What kind(s) of processing operation(s) do these terms denote?

4. A formula for a speciality breakfast sausage requires a maximum of 40% fat (F) and a minimum of 9% protein (P) in the final product. The production manager has three materials at his disposal: pork jowl (30% F, 15% P, 1.0$/kg); pork trimmings (55% F, 8% P, 0.2$/kg); and boneless picnic (20% P, 20% F, 1.80$/kg). If there are no further technological constraints, any two of these ingredients can be used to formulate the sausage. Based on the least cost formulation approach, determine which two materials should be used, their proportions in the final mix, and the cost of the formula. (Hint: calculate all three possible combinations for maximum permitted fat, then check the protein compliance and the price!)

5. What is myoglobin and how is it involved in appearance of fresh and cured meat products?

6. Why is sodium nitrite ($NaNO_2$) used in cured meat products? Why are many researchers actively working on projects aimed at developing a suitable substitute for use of nitrite in bacon?
7. Do meat grades give a good indication of nutritive value of a given meat cut? Why or why not?
8. What is aquaculture and how can it be developed into a significant human food resource?

**Suggested reference books—
Meat, Poultry, and Fish**

Aitken, A. Mackie, I. M., Merritt, J. H., and Windsor, M. L. 1982. *Fish Handling and Processing*. Aberdeen, Scotland: Ministry of Agriculture Fisheries and Food.

Connell, J. J. and Hardy, R. 1982. *Trends in Fish Utilization*. West Byfleet, Surrey, England: Fishing News (Books) Ltd.

Forrest, J. C., Aberle, E. D., Hedrick, H. B., Judge, M. D. and Merkel, R. A. 1975. *Principles of Meat Science*. San Francisco: W. H. Freeman & Co.

Kramlich, W. E., Pearson, A. M. and Tauber, F. W. 1973. *Processed Meats*. Westport, Conn.: AVI Publ. Co.

Kreuzer, R. 1974. *Fishery Products*. West Byfleet, Surrey, England: Fishing News (Books) Ltd.

Lawrie, R. A. 1974. *Meat Science*. Oxford, England: Pergamon Press.

Mountney, G. J. 1966. *Poultry Products Technology*. Westport, Conn.: AVI Publ. Co.

Price, J. F. and Schweigert, B. S. 1971. *The Science of Meat and Meat Products*. San Francisco: W. H. Freeman & Co.

Snyder, E. S. and Orr, H. L. *Poultry meat - Processing, Quality Factors, Yield*. Toronto: Ont. Department of Agriculture, Publ. No. 9.

8

Alcoholic and Non-Alcoholic Beverages

8.1 INDUSTRIAL BEVERAGE PRODUCTS

From the consumers' standpoint, liquid foods or beverages are the ultimate convenience food. Some typical representatives of this rather disparate grouping of industrial food products (fruit juices, pasteurized milk, carbonated beverages or whiskey) require no more effort than opening a bottle, a can or a carton. In other cases (tea, coffee, chocolate, frozen juice concentrate), a minimum of additional home preparation is needed to complement the achievements of the efficient industrial processor. In our modern society, virtually all the tedious labor that used to be associated with procurement of beverages—milking cows, roasting and grinding coffee beans, pressing apples to get the juice—has been handed over to industrial machinery.

Raw materials for manufacturing beverages and beverage ingredients come from all the main commodity groups discussed in this text, with the possible exception (at least in North America) of the meat group. Industrial production of some beverages is closely associated with other activities in a given commodity group. Thus, milk-type beverages are produced by dairy processors, fruit juices by the fruit-and-vegetable industry. In other cases (wine, beer, soft drinks), the beverage manufacturer has little in common with the fruit, cereal or sugar manufacturer except the same raw material.

The role that beverages play in modern society is several-fold. Perhaps first and foremost, beverages are consumed for their thirst-quenching properties. The traditional use of beverages as menu complements (coffee, tea, milk, breakfast juice) may be associated with this function, as well as with the rapidly emerging notion of a nutritious beverage. And finally, significant amounts of beverage products (alcoholic as well as non-alcoholic) are consumed as an inseparable part of various social activities. For the profit-oriented industrial beverage manufacturers, clear identification of intended users of their products is crucial for success in the intensely competitive beverage market. On the average, the human body requires about 1–1.5 L of dietary fluids daily; this can be translated to approximately 400–500 L of industrial beverages per capita per year. As shown in Table 8-1 the total consumption of industrial beverages has not changed drastically since 1960. Market success of one kind of beverage usually means a decrease in consumption of another kind. This is indicated especially by the increasing popularity of soft drinks and beer at the expense of coffee and milk products.

Technological principles used in production of the various beverages are as diverse as the major products themselves. Some of the processes were covered in the preceding text (see Chapter 4 for fruit juices and Chapter 6 for milk products). The remaining three major categories (alcoholic beverages, coffee and tea, and soft drinks) will be dealt with separately in this chapter.

8.2 ALCOHOLIC FERMENTATION

Brewing, wine-making, and manufacture of liquor products are based on the same fundamental principle—the conversion of a simple sugar into alcohol and carbon dioxide. The reaction proceeds according to

Table 8-1 Changing Patterns of Beverage Consumption in North America[a]

BEVERAGE	ANNUAL CONSUMPTION (LITERS PER CAPITA)		
	1960	1978	1990 (PROJECTED)
Milk	144	95	66
Coffee	136	91	63
Beer	57	88	105
Soft Drinks	47	137	190
Tea	21	45	62
Juice	10	15	19
Distilled Spirits	5	8	16
Wine	3	8	10
Total	423	487	531

[a]Source: Newsletter of National Dairy Council of Canada, Ottawa. 1979.

Equation 8–1, proposed first in 1815 by the famous French chemist Gay-Lussac, but known in principle to the ancient Egyptians several thousand years ago:

$$C_6H_{12}O_6 \xrightarrow{\text{yeast}} 2\ C_2H_5OH + 2\ CO_2 + \text{energy} \qquad \text{(Equation 8–1)}$$

In the production of alcoholic beverages, this reaction is carried out by selected strains of the yeast *Sacharomyces cerevisiae*. In Equation 4–2 we saw that the same biochemical reaction, anaerobic respiration, can proceed in fruit and vegetable tissues without the yeast and in the absence of oxygen. The substrate of the yeast fermentation (Equation 8–1) is glucose or other simple sugar. For wine fermentation, enough sugar is usually present in the juice of fruit used as raw material (most commonly the juice of grapes). In the production of beer the raw material is an aqueous extract of barley, often supplemented by other carbohydrate sources (corn, wheat, or simply sugar syrup). The barley must be malted (Section 5.6) since the fermentable sugar maltose must be first produced from the starch of the grain; this is accomplished by the enzymes activated in the malting process. Starch and/or sugar from almost any source is being utilized throughout the world for fermentation into the various liquor products characteristic of various countries. Some of the better known alcoholic products and the agricultural commodities used for their manufacture are listed in Table 8–2.

Similar to dairy cultures used in fermented dairy products (Section 6.4) the yeasts in alcoholic fermentation produce various aromatic compounds in addition to ethanol and CO_2, the principal end-products

Table 8–2 Plant Substrates for Fermentation of Distilled Alcoholic Products

PRODUCT	PLANT SOURCE	COUNTRY OF ORIGIN
Whiskey	Rye	Scotland
Bourbon	Corn	U.S.A.
Rum	Sugar Cane	Jamaica
Aquavit	Potato	Scandinavia
Tequila	Cactus	Mexico
Sake	Rice	Japan
Vodka	Sugar	U.S.S.R.
Slivovitz	Plums, Prunes	Czechoslovakia, Yugoslavia
Calvados	Apple	France
Cognac	Grapes	France
Shochu	Any fermentable carbohydrate	Japan

of their metabolism. Although these aromatic by-products including aldehydes, ketones, esters, organic acids or higher alcohols are produced in minute quantities, they, together with the compositional variations of raw materials used for fermentation, may play a major role in the sensory profile of some final products (wine, whiskey, etc.). In other instances, additional flavor ingredients are used to arrive at the characteristic final taste, especially in the case of beer and more so with liqueurs. The choice of ingredients as well as the raw materials and technological sequence of operations are the main differentiating factors in alcoholic fermentation of beer, wine, and liquors.

8.3 BEER AND THE BREWING PROCESS

Beer is the world's most popular alcoholic beverage. In North America it is projected that by 1990 the volume of beer consumed will be greater than that of milk, wine, and fruit juices combined (Table 8–1), and will be second only to soft drinks. In certain European countries (Germany, Czechoslovakia), beer has been a traditional component of the national food scene for centuries (Figure 8–1).

The principal component of beer, as in all beverage products, is water. Contrary to the production of fruit juices, wine or the dairy products in which water is an integral component of the raw material, pure well or tap water is used in the brewing process. Thus the quality of water available to the brewery is very important and must be standardized, especially for its mineral content. However, the common belief that only pure or deep well water can make a good beer, is one of the unproven traditions that is advantageously being used for marketing purposes. Some of the best Canadian or U.S. beers are made from regular city water supplies that have been properly standardized by demineralization or other modern water-treatment practices.

The initial ingredients in the brewing process shown schematically in Figure 8–2 are malted cereal grains (almost invariably malted barley or simply *malt*), and an additional source of carbohydrate (corn grits, wheat, rice, starch, or a sugar syrup, generally an *adjunct*). These essential sources of fermentable sugar are soaked in water in a processing step called *mashing*. Temperature control of this gentle cooking step is important in order to stimulate the optimum enzymatic activity needed to convert starches from cereals to simple fermentable sugars, and to extract the maximum soluble flavor compounds from the mash into the water. Several temperature changes are needed during the mashing step to accomplish the various biochemical and physical reactions desired; at the end of the mashing step, the temperature is raised to about 70°C to stop further enzymatic activity. The aqueous extract,

FIGURE 8-1. The world's oldest brewery in Weihenstephan, Germany, founded in 1040.

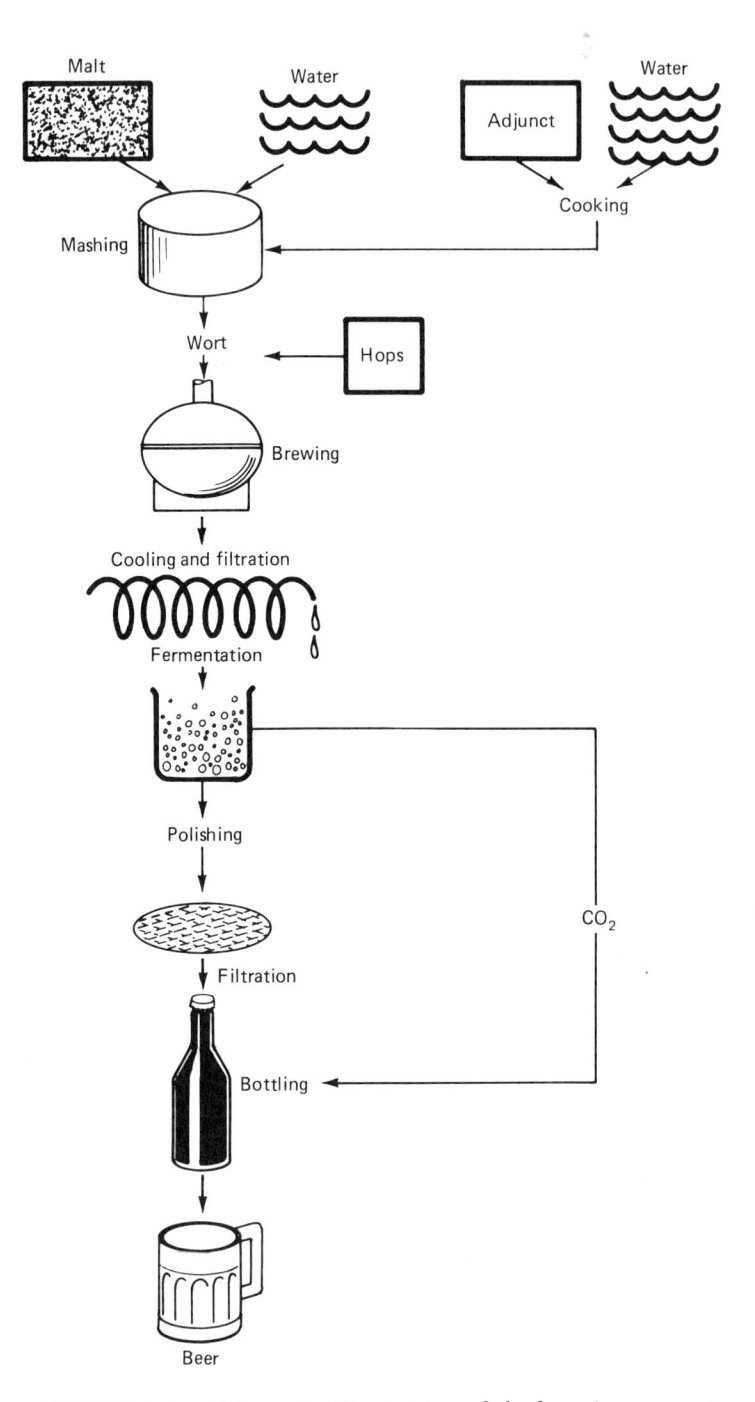

FIGURE 8–2. Schematic illustration of the brewing process.

referred to as *wort*, is separated from the spent grain by filtration through a perforated bottom of the cooking vessel or in a separate tank tradionally named *lauter tun*. The separated wort is boiled (brewed) in the brewing kettle, usually the centerpiece of a brewery (Figure 8-3).

One of the main purposes of this two to three-hour operation is to extract flavor compounds from the *hops*, the last major ingredient of beer manufacture added at this stage. The cone-like blossoms of the hops plant (*Humulus lupulus*, Figure 8-4) are used in the brewing process as a source of the characteristic bitterness of beer, coming from the compound *tannin* (found also in tea), as well as from essential oils and resins of the blossoms. Dried, whole blossoms of hops were used in the traditional brewhouses of Europe; nowadays hops is added often in pelleted form or as an extract of flavor compounds only.

In addition to proper flavor development, the boiling of wort is necessary to inactivate viable microorganisms that could later interfere with the yeast fermentation, to denature and precipitate proteins extracted from the mash, and to completely eliminate any residual enzymatic activity in the wort. The last two aspects are important for long shelf-life of the finished product, since the protein could cause cloud formation and the enzymes undesirable flavor changes. Upon

FIGURE 8-3. The traditional form of a brewing kettle.

FIGURE 8–4. The climbing vines of the hops plants (A) and the hops blossom cones used for the manufacture of beer (B). (Photographs courtesy of Sunny Hops Inc., Yakima, Washington.)

completion of the brew, the wort is cooled to the desired fermentation temperature (between 8–20°C depending on the type of beer being made) and transferred to fermentation tanks where brewer's yeasts are added.

The primary fermentation process lasts approximately nine days; thus, the fermentation cellars of a brewery must have adequate holding capacity. In the past this meant many large, expensive fermentation tanks. In modern beer production technology several batches of the brewed wort may be fermented in a single vertical "silo" similar to

FIGURE 8-5. The traditional beer fermentation tank (A) and the modern silo fermentors (B).

dairy silo tanks (Figure 8-5). Another recent innovation is the use of a more concentrated wort. After the fermentation process the concentrate is diluted with water to achieve the required alcohol and flavor content.

Although the term beer is used in the generic sense to describe all brewery products, there are several variations of beer fermentation that lead to different end results. According to the type of yeast and the fermentation temperature used, we may speak of lager or ale fermentation. Lager yeasts typically settle at the bottom of the fermentor kept at

Table 8-3   Main Categories of Brewery Products

PRODUCT	PRODUCTION CHARACTERISTICS
Lager Beer	Bottom yeast fermentation by *Sacharomyces ouvaris*, temp. 8–14°C
Ale Beer	Top yeast fermentation by *Sacharomyces cerevisiae*, temp. 16–21°C
Stout	Top or bottom fermentation, use of caramelized malt and/or caramel for dark color
Malt Liquor	Beer with alcoholic content above 6.0%
Light Beer	Low calorie (below 99/12oz.) and/or low alcohol (below 3.9% v/v)

8–14 °C, while the ale fermentation proceeds at 16–21°C and the yeast rises to the top. These and other differences among the various brewery products are briefly summarized in Table 8–3.

During the fermentation process sugars of the wort are converted to alcohol and CO_2 according to Equation 8–1. The volatile CO_2 is collected and eventually returned to the finished beer, which must first undergo further maturation (lagering at approximately 0–1°C), several filtration steps (to remove any yeast and other potential haze-forming substances), secondary fermentation (to further develop the alcohol content) or other special processes aimed at "polishing" the final desired quality. The properly aged beer (21 days to several months for some special products) is saturated with CO_2 and filled into bottles, cans, or kegs. The sealed bottles and cans are pasteurized in a conveyorized tunnel at approximately 65°C for several minutes to insure unrefrigerated shelf stability of several months. The *draught* (unpasteurized beer), typically sold in kegs and occasionally in bottles, must be kept refrigerated since its shelf life is much shorter.

The advance of science has eliminated much of the jealously kept but often poorly understood local variations of the brewing processes. Similarly, the developments in modern brewing technology and economic necessities of a fiercely competitive market have led to world-wide consolidation of the industry. In Canada as in the U.S., Australia and other major beer-drinking nations, there are usually only three to four major national brewing companies that own most of the operating plants throughout each country. While this may have diminished the taste differences among individual brands, it resulted in much more consistent quality and affordable price for this truly universal people's drink.

8.4 WINE AND LIQUOR PRODUCTS

Wine-making is a process as old as the recorded history of mankind. According to the Old Testament, "... Noah was the first tiller of the soil. He planted a vineyard; and he drank of the wine and became drunk...." (Genesis, 9, 20–21). The glory and the fall of the Roman empire is connected to enology (growing, making, and consumption of wine) in innumerable poems and historical accounts of contemporary writers. The phrase "in vino veritas" (Plinius) has become perhaps the most often repeated quote from Roman literature. In modern times, planting grape vines (the principal species used are *Vitis vinifera*, or *V. labrusca*) has been one of the important agricultural activities in newly settled lands such as America (New York, California), South Africa (the Cape region), or Australia (the Barossa and Hunter valleys).

The two most important wine-growing nations are France and Italy; however, Spain, Portugal and Germany are also known for significant production of high quality wine (Figure 8–6). In Canada, where the climate is unsuitable for traditional wine growing, progress has been made in recent years in establishing vineyards using new resistant hybrid varieties in the Niagara peninsula and interior B.C. regions. Wine production throughout the world is one of the most rapidly growing segments of the beverage industry since wine is becoming a universal part of world gastronomy.

In the generic sense, wine can be made by alcoholic fermentation of any fruit juice. Traditionally, wine means fermented grape juice. Products from other fruits (cherry, strawberry, orange, rose-hips) are manufactured in negligible quantities and the fruit must be identified on the label. The manufacturing process of converting grape juice to wine is simple. In the past, it required little more than squeezing the juice from the grapes and keeping it in suitable containers. The required yeast (*Sacharomyces cerevisiae* var. *elipsoideus*) is naturally found on the skin of grapes. However, to insure proper flavor development without possible interference by other undesirable ("wild") yeasts, the modern wine-making industry is using selected strains of the required yeast culture, which are added to the juice.

The modern industrial fermentation proceeds in carefully controlled conditions. Temperature control (approximately 25°C for red wines and not more than 15°C for white wines) is especially important as a substantial amount of heat (Equation 8–1) is evolved during the

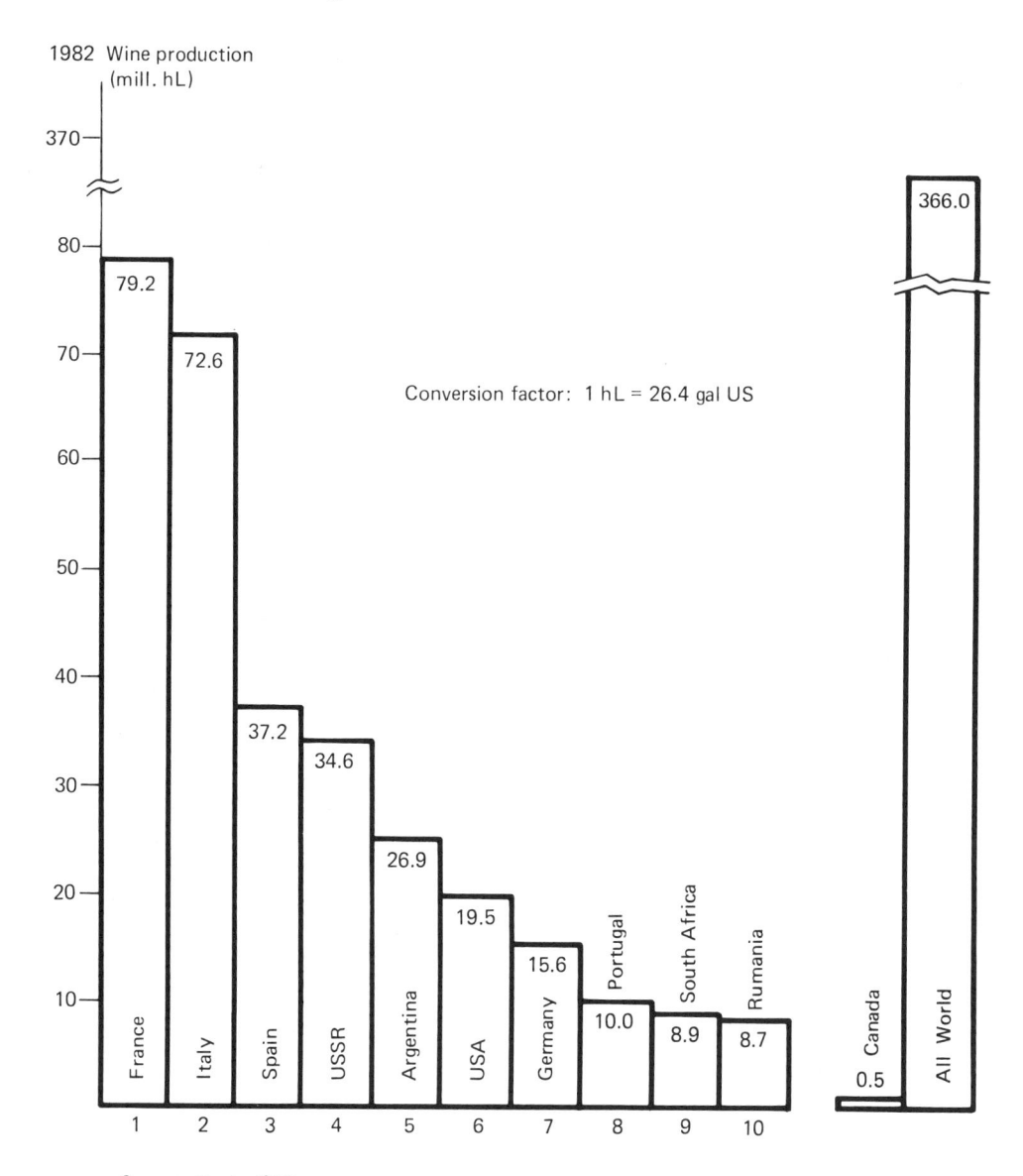

FIGURE 8–6. Diagrammatic illustration of world wine production. (Bulletin of the Office International de la Vigne et du Vin, 11 Rue Roquepine, 75008 Paris, France. Vol. 56:633, November 1983.)

fermentation process. The fermentors must be properly vented to remove the evolved CO_2 without allowing the outside air to contaminate the batch. After completion of the fermentation (several days to several weeks), the wine is transferred to settling vats for clarification or *racking* (repeated decanting to remove the slowly settling cloud of the freshly fermented young wine), and then to aging vats or casks.

The aging is a slow and complex process involving numerous biochemical reactions. The principal reactants are atmospheric oxygen and the various organic acids, carbohydrates, minerals, alcohols and other minute compounds coming from the grape components, the wood of the aging casks, or produced by the yeast metabolism. Generally speaking, the longer the aging (especially for red wines), the more balanced and delicate the flavor, and the more expensive the product. White wines usually don't improve in quality much beyond two years. For red wines, aging for five to ten years may produce some of the most sought after, rare delicacies depending on the region of production, the year (weather patterns during the growing season and at harvest have an important effect determining the exceptional, good or average years), and other often unpredictable factors.

Depending on the type of grapes used, sugar content of the freshly pressed grape juice, and the conduct of the fermentation, the finished product may be "dry" (no residual sugar left) or sweet. Some of the great medium sweet German white wines are made from grapes that were left on the vine for a late harvest (spatlese). The principal difference in making red or white wine is the fermentation of the juice only (white wine) or the complete crushed seed (*must*) containing skins and seeds in addition to the juice (red wines). An acceptable white wine may be made from juice of dark grapes, since the color of the red wine is extracted from the skins during the fermentation process. The current trend in wine consumption in the world is towards the less expensive white wines. Some of the traditional wines (named usually after the geographic region of origin) are listed in Table 8–4.

In North America, California is the most important wine-producing area. Its main advantages are the favorable climatic and soil conditions and wide selection of locally produced hybrid varieties, including the highly successful Zinfandel that typifies the success of the California wine industry. The traditional locally grown Canadian wines contain generally too much acid and not enough sugar for a choice product. Much of Canadian table wine is produced in modern plants located independently of the growing region and supplied by concentrated grape juice from other geographical areas. In fact, one of the most successful products of the non-traditional Canadian wine industry is the "Baby Duck" imitation of "champagne" where the CO_2 gas is supplied to an

188 Alcoholic and Non-Alcoholic Beverages

Table 8-4 Origin of Characteristic Wine Products

WINE	TYPE	ORIGIN
Bordeaux	Red or White	France (Bordeaux region)
Beaujolais	Red	France
Burgundy	Red	France (Burgundy region)
Chianti Ruffino	Red	Italy
Rhein	White	Germany, Rhine river valley
Moselle	White	Germany, Moselle river valley
Black Tower	White or Red	Germany, blended from several wines
Taylor	White or Red	U.S.A., New York region
Gallo	White or Red	U.S.A., California region
Corbans	White or Red	New Zealand, Auckland region
Zoenebloem	White or Red	South Africa, Cape region
Champagne	White, Sparkling	France, Champagne region

ordinary wine stock rather than produced by the traditional in-bottle fermentation process.

Fortified Wines

Complete fermentation of a typical grape juice will produce wine with 11–14% alcohol (v/v). If a higher alcohol content is desired, as in the various "aperitif" or "dessert" wines with up to 20% alcohol (sherry, vermouth, port), the fermenting wine is fortified with *brandy*, produced by distillation of regular wine. This increases the alcohol level and normally stops the fermentation before all the sugar is used up to achieve the desirable sweetness. The distilled wine in itself belongs to yet another class of alcoholic beverages—liquor products, of which the doubly distilled and gently heated white wine from the Cognac region of France is the noblest representative.

Liquor Products

The common characteristics of these beverages include their high alcohol content (often in the vicinity of 40% v/v or more), and their use as "social" drinks. In principle, there are two basic steps on which the various technologies from all over the world are based: alcoholic fermentation of a suitable, carbohydrate-containing liquid substrate, and distillation of the fermented substrate to increase its alcohol content. Some of the typical fermentable substrates are listed in Table 8–2. The most traditional ones, besides the grape juice, include aqueous extracts of grains (whiskey, bourbon, sake, vodka), potato (aquavit), sugar-cane juice (rum), fruit juices like plums (slivovitz), apricots (apricot brandy) and many others. In the distillation step the alcohol, as well as other volatile aroma constituents of the fermented liquid, will be preferentially boiled off with some of the water and later condensed. For a higher alcohol content a second distillation of the condensate may be necessary. The alcohol content in the final product is expressed in a unit specific to the alcoholic beverage industry—the degree of *proof*. In North America, a ° proof means ½% alcohol; a 100-proof product denotes 50% alcohol content by volume. This peculiar unit originated from the old custom of testing alcohol content in whiskey: the proof of appropriate composition, which was considered to be at least 50% alcohol, was obtained if gunpowder moistened with the product could still be ignited.

In addition to the required alcohol content, liquor products are differentiated primarily by their flavor. In some traditional products flavor differences may be associated with the geographical origin, as in the Scotch whiskies from the various mountain valleys. More often, the final desired flavor is obtained by blending distillates from various batches, aging in special processes (sometimes for years), or adding other aromatic compounds. This is especially true of liqueurs, where almost endless variations in flavor may be produced by selected herbs, spices, or some very common food ingredients such as milk, coffee, or cocoa. These often highly proprietary formulations are guarded by traditional manufacturers and imitated by others. Together with the ever-increasing variety of domestic and foreign wines being available, the alcoholic beverage market offers the consumer the widest imaginable selection of taste, prestige and price.

8.5 COFFEE AND TEA

Because of the similarities in production, processing and consumer use, it is convenient to consider these two popular beverage products together. Both tea and coffee are consumed purely for their flavor as a meal component or a social beverage, rarely (except iced tea) for their thirst quenching properties. Both are hot aqueous extracts (brews) of processed agricultural commodities grown in tropical or sub-tropical regions of the world.

Coffee comes from fruits ("beans", or more correctly, cherries) of a perennial coffee tree *Coffea arabica* and several other species of the genus *Coffea*; tea is produced from leaves of a perennial tea bush, *Camellia sinensis* (var. *sinensis* or *assamica*). In both cases, the desired flavors are developed in several post-harvest processing steps which are different for each product.

Coffee beans are dried first to decrease their moisture content for prevention of spoilage in storage and transportation. The most important step in coffee bean processing is *roasting*, or dry heat application, which results in extensive chemical changes of the coffee bean constituents. The temperature of roasting is high, often exceeding 200°C, and this results in heat breakdown (pyrolysis) of the carbohydrates, leading to production of numerous color, flavor and aroma compounds. Currently available scientific techniques identified over 300 chemical compounds contributing to coffee flavor, most of them present in minute quantities. The most important type of compounds found in roasted coffee include aldehydes, ketones, esters, alcohols, hydrocarbons, pyrazines, phenols and other organic volatiles. The exact contribution of these compounds to the delicate flavor of properly roasted coffee is unknown. As a result, coffee is almost impossible to duplicate by the manufacturers of artificial flavors. Roasting produces many other chemical changes such as destruction of thiamin, changes in the protein and lipid components, and degradation of chlorogenic acid and other phenolic compounds. The chlorogenic acid destruction is especially important since it results in the elimination of bitterness present in green beans.

The technological nature of *tea* processing is somewhat different. Freshly harvested tea leaves are left to wither for less than a day. The leaves are then gently crushed to damage the leaf cells just enough to release the juices and the enzymes contained in them. The withering and crushing are vital steps for the main flavor-producing operation called the *fermentation*, during which biochemical changes (predominantly enzyme-catalyzed oxidation of various compounds including catechol, flavonols and tannins) produce the desirable flavor. Careful control of this delicate processing step is necessary; one of the indicators used is the color change from green to copper-red, caused by the tannin oxidation. When the fermentation is considered complete (usually in three to five hours), the leaves are "*fired*" (dried) at approximately 90°C to stop all enzymatic activity, darken the product (*black tea*) and remove moisture to insure adequate shelf-stability. The traditional description of the process as fermentation is obviously a misnomer since no microbial activity is involved. In some cases, unfermented (*green tea*) or semi-fermented (*oolong tea*) products are manufactured depending on the local market preferences.

The flavor chemistry of the finished tea is just as complex and even less understood than that of coffee. Of the almost 400 compounds contributing to the characteristic tea flavor, only about 140 have been positively identified. The main compounds include organic acids, aldehydes, theaflavins, thearubigins, alcohols, esters, terpenes, lactones and other degradation products of enzymatic oxidation of tea tannins or amino acid breakdown. It appears that the aroma compounds originate mainly as a result of the biochemical processes during fermentation, although some additional compounds are developed by heat in the firing process and some undesirable volatiles are removed.

To prepare a coffee or a tea beverage, the consumer must extract the flavor compounds from the processed materials by hot water. Worldwide, there exist numerous modifications of this seemingly simple operation, resulting in distinctly different final products which are often characteristic of a particular culinary tradition. Various coffee serving styles are among the most interesting traditions in several European cuisines, as shown in Table 8–5. For added consumer convenience, an industrial process for flavor extraction from coffee or tea has been developed and is being used for production of the soluble (*instant*) coffee and tea products. The process is based on the same principle used at home where the insoluble components of ground coffee beans or tea leaves are also discarded. For marketing purposes, however, the extracted coffee or tea solubles must be dried (usually by spray drying or freeze drying, see Chapter 10) and this is where some of the most volatile aroma compounds may be lost. Further precautions must be taken in packaging these delicate products, since their main quality attribute, the taste, may change in prolonged storage as a result of oxidative changes or volatilization. Thus, most soluble coffee and tea

Table 8–5 Traditional Coffee-Serving Styles

STYLE	CHARACTERISTICS
Turkish	Hot water poured onto ground coffee in a metal decanter; extract decanted into cups
Viennese	Black coffee served with whipped cream on top
Espresso	Percolated extract of steamed coffee beans
Café-au-lait	Coffee served with the rich addition of milk or cream (up to 1:1 ratio)
Irish Coffee	Black coffee served with whiskey and cream

products are packaged under nitrogen atmosphere in sealed gas-impermeable jars.

Drinking tea or coffee is one of the social habits of many Occidental as well as Oriental cultures. The widespread popularity of these beverages may be related to their refreshing stimulatory effect that comes in part from one of their more controversial natural chemical compounds, *caffeine*. Heavy drinking of either tea or coffee may result in relatively large amounts of caffeine being ingested. There is approximately 80 mg of caffeine in a 120 ml cup of coffee, and about 40 mg in a cup of tea. In the U.S., the average daily intake of caffeine from all dietary sources (these include also cocoa and some soft drinks, see Section 8.6) is about 200 mg/adult. Decaffeinated coffee products (instant or regular) are available for those who may want to avoid caffeine for various personal reasons. However, the decaffeination, which is achieved by a special selective extraction process, will also remove some flavor components so that the resulting product may have a somewhat different characteristic.

In recent years, caffeine came under attack as a naturally occurring compound that may have undesireable health effects if ingested in large doses. In the purified form, caffeine is being used as an active compound in various prescription and over-the-counter drugs such as diuretics, weight control aids or cardiac stimulants. The pharmacologically active dose used in these medical preparations is about 200 mg. In various epidemiological, psychological, gynecological and other medical studies, the effects of caffeine, or more generally, coffee-drinking, have been evaluated in relationship to cardiovascular disease, blood cholesterol levels, pregnancy complications and cancerogenicity. There appears to be inconclusive and still sketchy evidence of possible undesirable effects at high levels of ingestion. In contrast, many other studies seem to indicate that in moderation, coffee and tea can be enjoyed without any adverse effects.

8.6 SOFT DRINKS

In terms of consumption statistics, the non-alcoholic carbonated beverages as a group are the fastest growing commodity on the food market (Figure 8–7). In the U.S., the consumption of soft drinks increased by some 300 percent between the years 1960 and 1979. Expressed as the number of 12-oz. containers per person per year this translates to a startling average of over 400 cans for every man, woman and child. The success of the soft drink industry, resulting from the ever-increasing consumer demand as well as from the relative simplicity of technology, is one of the major factors affecting all beverage manufacturers (Table 8–1).

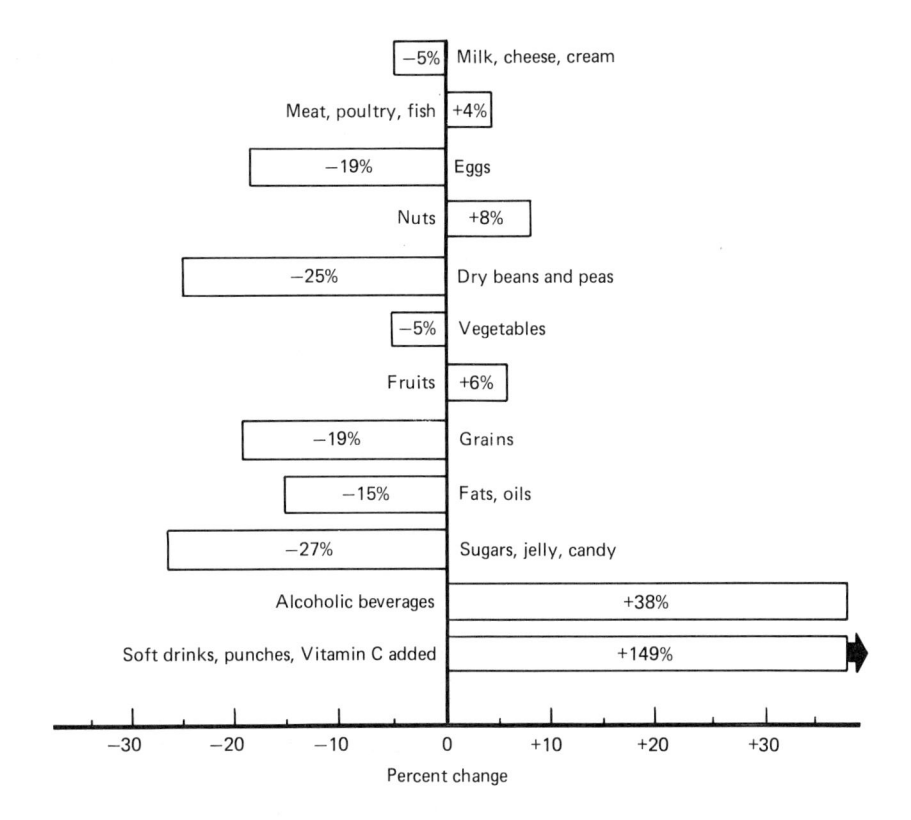

FIGURE 8-7. Changes in food consumption in the United States. (Dorothy Wenck. *Supermarket Nutrition*, 1981. Reprinted by permission of Reston Publishing Co., Inc., a Prentice-Hall Company, 11480 Sunset Hills Rd., Reston, VA 22090.)

The origins of carbonated soft drinks may be dated back to the popularity of mineral waters in the Roman empire. The artifical carbonation of water by sodium bicarbonate (thus "soda water" and later "soda pop") was invented in 18th century England. In the U.S., one of the milestones for the soft drink industry was the development of Dr. Pepper (1885) and Coca-Cola (1886) beverages, which contain extracts of the kola nut as flavoring. The principal ingredients of pop drinks today are water, sugar or other sweetener, artificial or natural flavorings, acid, CO_2 gas, and often a coloring and a preservative.

Apart from water, sugar is by far the most abundant ingredient. Its relatively high level in most of the regular drinks (Table 8–6) is causing some concern among nutritionists, particularly in conjunction with the current consumption pattern (Table 8–1). An alternative for the modern calorie-conscious consumer, the low calorie "diet" drinks, are now becoming increasingly popular. The necessary sweetness is provided

Table 8-6 Main Compositional Characteristics[a] of Carbonated Soft Drinks

DRINK TYPE	ACIDITY (pH)	SUGAR CONTENT (%)	CAFFEINE CONTENT mg/100 ml
Regular colas	2.6	10–13	12
Diet colas, decaffeinated	2.6	0	Trace
Root beer, ginger ale	4.0	7–11	Trace
Orange, lemon-lime, other artificial fruit flavors	3.0–3.5	10–14	0

[a]Data from "The facts about soft drinks", published by Canadian Soft Drink Ass'n., Toronto, 1984, and other sources.

by an artificial sweetening agent such as the traditional saccharin or the recently discovered aspartame (a protein-like compound consisting of residues of two amino acids, phenylalanine and aspartic acid). These compounds give a mouth sensation similar to that of the common sugars, but, because the sweetening power of the artificial ("non-nutritive") sweeteners is many times greater than an equal amount of sugar, the amounts needed in a diet drink are very small. While the flavor impact of the non-nutritive sweeteners is not identical to that of sugars, the virtual elimination of any food energy from the diet drinks often outweighs the occasional flavor defects. To minimize flavor changes from the traditional sugar-containing products, a naturally occurring sugar fructose (found predominantly in honey and some fruits) may be sometimes used to produce "calorie-reduced" drinks. Since the fructose is about 1.5 times sweeter than the regular table sugar, fructose-containing products will have approximately ⅓ less calories than the regular drink while still maintaining the same sweetness and flavor profile.

Another disputed ingredient of soft drinks is caffeine (Section 8.5) found primarily in cola-type beverages (Table 8-6). The typical level is about one-fifth of that found in a cup of coffee. Caffeine is a natural component of the kola nut, but extra caffeine may be added in the soft-drink manufacturing process. Caffeine-free cola drinks have been recently introduced by the major manufacturers in response to the growing concern over caffeine consumption, particularly by children.

The technology of soft drink manufacture is simple. The success of a soft drink is determined almost entirely by its flavor; the production of the flavor base (concentrate) is the most important aspect of the process. The concentrate, containing all the required ingredients except water and the CO_2 gas, is usually produced in a central plant, often under precautions of high secrecy. Supplies of the concentrate are delivered to numerous bottling plants throughout the country where properly treated water and CO_2 gas are added. The carbonation is an interesting process based on the solubility characteristics of CO_2, typical of all gases. As the solubility increases with increasing pressure and decreasing temperature, the soft drink carbonators are refrigerated pressure vessels in which the beverage is contacted with a CO_2 gas from a suitable supply—usually a gas cylinder. Upon dissolution, the CO_2 reacts with water according to Equation 8-2 to form a weak carbonic acid:

$$H_2O + CO_2 \longrightarrow H_2CO_3 \quad \text{(Equation 8-2)}$$

The carbonic acid alone would not be effective in producing the desired acidity in the final product (typically pH 2.6–3.2), and thus citric acid (for fruit-flavored drinks) or phosphoric acid (for colas and other non-fruit flavors) are added. The high acidity of the soft drinks is necessary for their thirst-quenching properties and also as a preservative, since carbonated beverages are not pasteurized or otherwise heat-treated. Chemical preservatives such as sodium benzoate may also be used to insure sufficient shelf-stability, particularly in some of the relatively low-acid fruit-flavored products.

Soft drinks are often attacked as empty-calorie junk food. It should be remembered though that beverages in general, and soft drinks in particular, are consumed mainly to quench thirst—or, more scientifically, to satisfy the physiological needs of an active body. In this regard, they play the same role as a refreshing glass of ice-cool water; the cola drinks may have the added mildly stimulatory effect of caffeine which may be appreciated especially after a strenuous physical activity. The market success of soft drink products is unquestionable, and the consumer willing to buy is an active partner of the industrial processor in this economically important area of food processing.

8.7 NUTRITIONAL AND MEDICAL IMPLICATIONS OF INDUSTRIAL BEVERAGES

As a source of important nutrients, the beverages covered in this chapter are of minimal significance. While fruit juices (Chapter 4) and milk (Chapter 6) are valuable as nutritious drinks, alcoholic beverages, coffee and tea, and soft drinks are consumed primarily for other than nutritional reasons. The benefits of these beverages are in their thirst-quenching properties, their value as social drinks, often enhancing the pleasure of eating and the efficiency of digestion, and for some possible medical reasons. The absence of identifiable microbiological hazards of thirst-quenching beverages due to the nature of the products and/or the manufacturing processes, may be best appreciated when traveling in countries lacking adequate sanitary water supplies. Beer and wine were used for medicinal purposes in ancient Egypt. Even today, some physicians prescribe certain red wines as a source of iron; other recommend wine as a sedative, against hypertension or a nutritional supplement for diabetics. Consumption of alcohol in small amounts has been recommended for therapeutic properties related to cholesterol levels in blood and for improved digestion of fats in the diet. The controversial subject of caffeine in tea and coffee has been briefly discussed in section 8.5. About 1,000 prescription drugs and 2,000 over-the-counter pharmaceutical products, however, also contain caffeine. The effect of caffeine as a cerebral stimulant has been associated with increased work capacity by students and others. The availability of soft drinks and especially their diet varieties (containing no sugar) has been positively correlated with mental health of certain obese patients.

As with many other foods, excessive consumption of alcoholic and non-alcoholic beverages may result in severe undesirable effects. Possibly the least significant one is the relatively high energy value of soft drinks and of alcoholic beverages with higher alcohol content. In Table 1–7 we saw that alcohol has a high energy value (7 Kcal/g as compared to 9 Kcal/g of fat and 4 Kcal/g of carbohydrate). A comparison of energy content of some alcoholic and non-alcoholic beverages is in Table 8–7. A much more serious sociological phenomenon is the

Table 8–7   Energy Content of Selected Alcoholic and Non-Alcoholic Beverages[a]

BEVERAGE	ENERGY CONTENT/100 ml[b] KCAL	kJ
Orange juice, fresh	45	190
Cola-type carbonated beverage	40	170
Diet cola	<2	<10
Beer (4.5% alcohol v/v)	42	175
Wine (12.2% alcohol v/v)	85	360
Whiskey (90 proof)	265	1,115

[a]Source: "Nutritive value of American foods" by C. F. Adams, Agric. Handbook No 456, USDA, Washington, D.C., and other sources.
[b]For definitions of Kcal and kJ see Table 1–7.

behavioral aspect of alcohol consumption in large amounts. The drunk driver is rapidly becoming one of the primary causes of premature death in North America, while chronic alcoholism has long been one of the most serious social diseases all over the world. The suggestions of toxicological complications resulting from high doses of caffeine are also a cause of concern, particularly regarding possible effects of chronic exposure to caffeine in children and pregnant women.

There is little doubt that alcoholic and non-alcoholic beverages are important food products. The enormous economic impact of the beverage industry in terms of consumer spending and corporate taxes is highly beneficial to modern society. The undesirable side-effects of some of the final products are not the fault of the product but of the indiscriminate user. In moderation, a glass of wine, a can of soda pop, or a cup of tea can make life much more enjoyable.

8.8 CONTROL QUESTIONS

1. Why is yeast necessary for an alcoholic fermentation? Can bacterial cultures be used for the same purpose? Can alcoholic fermentation proceed without a yeast? Is the tea fermentation process an alcoholic fermentation?
2. Explain the difference between lager beer, ale, stout, and malt liquor.
3. What is the purpose of an adjunct in the beer-making process? Could cheese whey be used as an adjunct? Why or why not? What is the purpose of hops?
4. An Italian white wine may contain typically 10% alcohol v/v. Calculate the approximate w/w concentration of alcohol in this wine (the density of alcohol can be taken as 0.8 g/ml). Using your result and the stoichiometric alcoholic fermentation equation, can you estimate the approximate sugar content of the grape juice from which the wine was made?
5. What is the food energy content of a 1 oz (approx. 30 ml) glass of Canadian whiskey conforming to Canadian specifications (the maximum alcohol content of Canadian liquors may not exceed 80° proof)?
6. What are the reasons for using heat during processing of coffee beans and tea leaves into commercial products? Are the effects of the heat processing the same for both products? Explain the changes occurring as a result of the heat application.
7. What is the role of carbon dioxide (CO_2) in beer, sparkling wine

or soft drinks? Why is the CO_2 lost if these drinks are left standing at room temperature for too long?

**Suggested reference books—
Brewing, Wine-Making, Coffee, Tea, Soft Drinks:**

Amerine, M. A. 1981. *Wine Production Technology in the United States.* Washington, D.C.: Amer. Chem. Soc.

Amerine, M. A. and Roessler, E. B. 1976. *Wines-Their Sensory Evaluation.* San Francisco: W.H. Freeman & Co.

Amerine, M. A., Berg, H. W., Kunkee, R. E., Ough, C. S., Singleton, V. L., and Webb, A. D. 1980. *The Technology of Wine Making.* Westport, Conn.: AVI Publ. Co.

Lee, F. A. 1983. *Basic Food Chemistry* (pp. 211-229, 281-327). Westport, Conn.: AVI Publ. Co.

Pollock, J. R. A. 1979. *Brewing Science.* New York: Acad. Press.

Rose, A. H. 1977. *Alcoholic Beverages.* New York: Acad. Press.

Vine, R. P. 1981. *Commercial Winemaking.* Westport, Conn.: AVI Publ. Co.

A Toast to Ontario Wines. Toronto, Ontario: Wine Council of Ontario.

Woodroof, J. G., and Phillips, G. F. 1981. *Beverages-Carbonated and Noncarbonated.* Westport, Conn.: AVI Publ. Co.

9

Food Additives and Food Ingredients

9.1 WHAT ARE FOOD ADDITIVES?

Of all the foods and food substances handled by the food processing industry today, food additives are undoubtedly the most controversial group. Many consumers are becoming increasingly skeptical about "chemicalization" of our food supply, and food additives are singled out as the perceived reason. To a large degree, this is due to the labeling regulations requiring a list of ingredients on all composite foods. Thus, many food items that are manufactured by mixing various ingredients carry labels full of chemical-sounding compounds which have been used in the manufacturing process. As an example, Table 9–1 shows a list of ingredients found on a jar of a cheese spread. Of all the items listed, only a few are true food additives according to Canadian or U.S. food laws. The public, however, seems to consider anything having a chemical name to be an additive.

The public attitude towards food additives may be best characterized by a comment from an anonymous consumer included in a survey that the Canadian Health Protection Branch[1] conducted in 1979: "If it was not in the food product to start with or would not have been present naturally but was added at any point in the production of the food, it is a food additive." If this were to be the regulatory approach, then essentially all food items used in manufacturing of other food items would become additives. Skim milk powder used in sausage would

---

[1] Publication 48, Health and Welfare Canada, Ottawa, 1980.

Table 9-1 List of Ingredients[a] in a Cheese Spread Product

INGREDIENT	STATUS AS ADDITIVE CANADA	U.S.A.
Milk solids	No	No
Water	No	No
Concentrated and/or powdered whey	No	No
Whey butter and/or whey butter oil	No	No
Bacterial culture	No	No
Sodium phosphate	Yes	Yes
Salt	No	Yes
Mustard flour	No	No
Rennet and/or pepsin and/or microbial enzyme	Yes	Yes
Calcium chloride	Yes	Yes
Flavorings	No	Yes
Spices	No	Yes
Sorbic acid	Yes	Yes
Color	Yes	Yes

[a] As declared on the label.

have to be considered an additive. Egg whites, spices, and many other ingredients would have to be subjected to the same rigorous testing procedures that are needed for true food additives to insure that no questionable materials are used by the food processor. Clearly, such an approach is unrealistic.

The term "food additive" has become widely used—and abused—as a result of the 1958 Food Additives Amendment to the 1938 Food, Drug and Cosmetics Act of the U.S. The legalistic aspects of the food additives regulation are complex and confusing for the consumers, who may not be aware of the technological reasons for using added substances (whether or not they are legally considered additives) in processed foods, and the testing required before a new substance is allowed as an additive.

9.2 FUNCTIONAL PROPERTIES

When food processing began moving from farmers' back yards to industrial plants, the economics of modern technology and market competition became important. A better looking or better tasting product, or product of equal quality that could be produced more cheaply, would likely attract the largest share of the market. A bologna sausage that would show free fat or gelatinous pockets would not be tolerated by consumers if the same product, manufactured by a competitor, would be free of these defects. Gritty ice-cream with large ice-crystals is considered inferior if a smooth product is also available. Green oranges cannot be sold even though they may be just as ripe as orange ones. Several other situations related to consumer choices are illustrated by the short questionnaire in Table 9–2. If your answer to these questions is "yes," then you too are one of those whom the food processing industry is trying to satisfy by using food additives.

Most manufactured food products contain more than one component. The reasons for using the individual components may vary greatly. In general, they are related to desired final flavor, texture, nutrient content (including water), or other quality aspects. When different types of meat are used in a sausage to produce the optimum flavor, the fat contained in these meat ingredients may or may not be easily emulsified with the meat protein, depending on the manufacturing process. If a proper meat binder (Chapter 7) such as skim milk powder, wheat flour or other approved food substance is added, the desired product quality can be achieved and maintained. The selection of the most appropriate binder in this example will be based on its functional properties or *functionality* in a given food system. Typical functional properties of interest to food processors include solubility, water binding, fat binding, emulsification, acidity control (buffering) and several other technological properties listed in Table 9–3. In a more general concept, the ability to provide optimum color, flavor, sweetness or longer shelf life can be also considered a functional property.

From the standpoint of a food processor, all components of manufactured food can be considered functional ingredients. In ice-cream, milkfat provides the smoothness of the background dairy flavor and, together with the dairy protein, the "body" (texture); sugar is added for sweetness, and it also contributes to the texture; and flavor-

Table 9–2 A Food Quality Questionnaire
(Answer These Questions According to Your Own Preference)

	Yes	No
1. Do you prefer a chocolate-flavored milk in which the chocolate does not settle so that you don't have to shake it before drinking?		
2. Do you get upset if the salt does not pour from the salt shaker because the grains are stuck together?		
3. Do you like your cake batters to rise reliably for your home-baked products?		
4. Are you happy that your bread does not get moldy on your home shelf?		
5. Do you want nutrients which are inadvertently removed during processing to be returned to the food?		

Table 9-3   Some Functional Properties of Importance to Industrial Food Processors

PROPERTY	DESIRED FUNCTION	EXAMPLE OF INGREDIENT
Water Binding	Prevent water from leaking from solid products	Protein
Emulsification	Form stable emulsions of water and fat	Lecithin, Monoglycerides
Whippability	Form stable foam upon aeration	Egg white protein
pH Control (buffering)	Maintain proper pH in mixed food systems	Phosphates
Heat Gellation (coagulability)	Form gel upon heating	Gelatin, egg white, soy protein
Thickening	Increase viscosity of liquid products	Starches, carboxymethylcellulose
Oxygen Binding	React with oxygen to protect foods from fat oxidation	Butylated hydroxyanisole (BHA)
Firming	Maintain proper firmness upon heating (canned vegetables)	Alum

ing (usually artifical because of the lower cost) is used to satisfy consumer preferences. Emulsifiers and stabilizers are important functional ingredients for maintaining the delicate emulsion balance during processing, controlling the size of the ice crystals formed upon freezing, and insuring the lightness of the product by maintaining the proper viscosity needed for air incorporation. Many of the ice-cream ingredients come from natural sources, including cow's milk, sugar beet or cane, sea-weeds (stabilizers such as carrageenan) or soybeans (emulsifers such as lecithin). Other required compounds are synthesized in a laboratory from chemicals similar to those used by "Mother Nature" (artificial vanilla and other flavors, artifical food colorants). All these ingredients are used in a batch of ice-cream because of their functionality which will contribute to the best possible quality at the lowest possible cost.

9.3   FOOD INGREDIENTS

Many products of the modern, diversified food processing industry are manufactured not only for home use, but also for industrial use by other food processors. Examples of predominantly industrial-use products described elsewhere in this book include concentrated proteins from oilseed processing, special dairy products such as modified whey powders, caseinates, or lactose, mechanically separated ("deboned") meat paste, dried onion powder, and many other products. In some instances (wheat flour, herbs, spices), the primary market for a specialized segment of the food processing sector is the industrial food processor. In yet another situation typified by the egg-processing industry, industrial ingredient products (primarily dried egg powder and dried egg whites), may be significantly different from the products destined for the consumer (essentially fresh eggs sorted according to their size and checked for quality by *candling* or visual inspection above a powerful light source). This last example also illustrates the differences between quality requirements of retail and industrial markets. Fresh eggs are sold primarily for their nutrient content, while functional properties (whippability and heat coagulability) are the main required quality attributes of industrial egg-white powder. Maintaining and maximizing the functionality in this case dictates the available processing technology (spray drying or freeze-drying—see Chapter 10).

Many products of the food ingredient industry are used for functionality. Traditionally, spices contribute unique flavors, starches gelatinize upon heating, dairy proteins are easily soluble in water, and egg-yolk powder provides superior emulsification for manufacture of mayonnaise. Rapidly advancing food science research has shown that in many cases, similar functionality can be provided in some products by other innovative food ingredients at much lower cost. Modified soybean protein products provide solubility, heat coagulability and whippability equal to or better than the expensive dairy or egg-white proteins which had to be used in many food applications separately. Similarly, whey protein ingredients can be substituted for egg whites in some cases requiring whippability (Figure 9-1), or hydrogenated vegetable oils (Chapter 5) may be used instead of butterfat to produce an acceptable imitation ice cream ("mellorine"). Some ingredients are used for their flavor as well as color; examples include strawberry jam in yogurt, chocolate syrup in ice-cream, cheese cubes in certain types of bologna, or mustard in prepared salads and relishes.

Rapid progress in many areas of food research has advanced our understanding of the physicochemical mechanisms involved in the functionality of many traditional food ingredients. Similarly, recent concerns with nutritional aspects of processed foods have led to process or product modifications that have resulted in new demands for improved functional ingredients. Some traditional ingredients are becoming too expensive for the functional benefit they provide, especially if only one of their components is needed. If a specific functional property is desired, it is often cheaper and technologically superior to

FIGURE 9-1. Whey protein foam (prepared from specially processed cheese whey) resembling whipped egg white.

use an isolated, chemically well-defined single substance rather than a complicated natural mixture of many chemical substances of which only one performs the needed task. In short, it may be more advantageous to use a food additive rather than a traditional food ingredient.

9.4 FOOD ADDITIVES

To ensure proper control, food additives are legally defined in both U.S. and Canadian food legislation, and their use is strictly regulated. Table 9–4 lists the major categories of food additives and examples of their use. Appendix III gives a more detailed list of commonly used ad-

Table 9–4 Main Categories of Food Additives[a]

CATEGORY	EXAMPLE ADDITIVE	USED IN
Anticaking agents	Calcium silicate	Salt mixtures
Dough conditioners, bleaching and maturation agents	Potassium bromate	Flour, bread
Coloring agents	Annato yellow	Cheddar cheese
Emulsifiers, gelling and thickening agents	Carrageenan	Ice cream
Food enzymes	Papain	Meat tenderization
Glazing and polishing agents	Maganesium silicate	Confectionery products
Firming agents	Calcium chloride	Canned vegetables
Non-nutritive sweeteners	Aspartame	Diet beverages
Acidulants	Citric acid	Fruit and vegetable products
Preservatives and antioxidants	Sodium nitrite, sorbic acid, butylated hydroxyanisole (BHA)	Prepared meats, jams, fatty foods
Sequestering agents	Calcium phosphate	Ice cream mix
Starch modifying agents	Sodium carbonate	Starch
Yeast foods	Zinc sulphate	Beer
Carrier or extraction solvents	Ethyl alcohol	Spice extracts
Miscellaneous (antifoaming agents, humectants, pressure-dispensing agents, whipping agents)	Nitrogen; sorbitol	Whipped toppings; marshmallows

[a] Based on classification of Canadian Food and Drug Act and Regulations, Ottawa.

ditive compounds and explains their functions. We can see that a majority of the additives are used for a specific technological reason; only one of the fourteen major classes of food additives are chemical preservatives used to prolong shelf-life of some processed foods.

Many of the food additives are natural compounds isolated from plants or animals; the best known examples include the emulsifier lecithin from soybeans, the cheese-producing enzyme rennet from calves' stomachs (Chapter 6), or the yellow annato cheese color coming from a tropical tree. Other additives are simple organic or inorganic compounds (calcium silicate, aluminum sulphate, citric acid, sodium bicarbonate) found in nature or synthesized in a laboratory. Some of the synthetic colors or flavors are prepared in laboratories from various organic compounds that are found in the natural sources of these functional additives (in Canada synthetic flavors are in fact considered an ingredient rather than an additive and they are regulated under a separate section of the Food and Drug Act). However, the fact that many additives come from a chemical laboratory does not make them any different from those same or similar compounds found in nature. The main difference is that any food additive, before it is permitted to be used in foods, must be thoroughly tested for its safety. If many of the traditional natural ingredients or foods were put to the rigors of the current additive testing procedures, they most certainly would not be allowed as foods. The best examples are wine, beer and liquor products (their alcohol content in the doses required for the tests would probably kill most of the test animals), coffee and tea due to their caffeine content, or potatoes because of the minute amounts of solanine. According to the current government regulations, the burden of proof of the additives' safety lies with the industry. However, the regulatory agencies may carry out extensive laboratory testing programs to confirm the data supplied by the industry.

A major difference between U.S. and Canada exists in government regulatory approach concerning the use of food additives. In the Canadian "Food and Drug Act," additives are defined as "*any substance the use of which results or may reasonably be expected to result in it or its by-products becoming a part of or affecting the characteristics of a food.*" By virtue of separate regulations governing the food use of each category of ingredients such as herbs, spices, seasonings, flavoring preparations, salt, sugar, starch, vitamins, minerals and amino acids, these components are excluded from the category of food additives. Likewise, agricultural chemicals, veterinary drugs and food packaging materials are covered separately by the Food and Drug Act of Canada.

The American legal definition is much more complex. It denotes food additives as "*any substance the intended use of which results or may reasonably be expected to result, directly or indirectly in its becom-*

ing a component or otherwise affecting the characteristics of any food, including any substance intended for use in producing, manufacturing, packing, processing, preparing, treating, packaging, transporting or holding the food." The cornerstone of U.S. legislation is a provision for approval of food substances according to specific safety evaluation procedures *"... if such substance is not generally recognized among experts qualified by scientific training and experience to evaluate its safety, as having been adequately shown through scientific procedures (or, in the case of a substance used in food prior to January 1, 1958 through either scientific procedures or experience based on common use in food) to be safe under the conditions of its intended use."* Based on this definition found in the 1958 Food Additive Amendment to the Food, Drug and Cosmetics Act (1938), special regulatory procedures were developed which created three categories of food substances. These include: 1) ingredients and other compounds that have been used before 1958 and that are "generally recognized as safe" (GRAS); 2) substances considered unsafe according to the well known *Delaney Clause* contained in the 1958 amendment (*"no additive shall be deemed to be safe if it is found to induce cancer when ingested by man or animal, or if it is found, after tests which are appropriate for the evaluation of the safety of food additives, to induce cancer in man or animal"*); and 3) all other non-traditional food substances that have to be tested case by case before they are allowed in processed foods.

Following Congressional approval of the 1958 Food Additives' Amendments, a lengthy list of substances which were in general use before 1958 (the "GRAS list") has been established by the FDA. This list contains several thousands of products including salt, sugar, corn syrup, and other frequently used ingredients that constitute the bulk of the "chemical additives" consumed in the U.S. Because of the "grandfather clause" approach to establishing the GRAS list, only very few of the GRAS substances underwent the rigorous testing required of all newly approved additives. Thus it is not surprising that several compounds (such as the non-nutritive sweeteners cyclamate and saccharin) were removed from the list after new—even though controversial—scientific evidence became available. The GRAS list is now under constant revision and a number of substances were deleted from or added to the original list.

The different approaches to regulatory control of food additives in the U.S. and Canada have resulted in several paradoxical situations, the most notorious case being the banning of the food color, "Red Dye No. 2" in the U.S. (but not in Canada), and the prohibition of "Red No. 40" in Canada (but not in the U.S.). The common underlying cause for this and other discrepancies is the wording of the Delaney Clause, which does not take into consideration the tolerance levels of the human body. According to this law, any substance that may have caused cancer in experimental animals anywhere, under any conditions (including extremely high dosage or intravenous feeding), is not allowed in food in the U.S. A recent example concerning the artificial sweetener saccharin may be used to illustrate the point. Saccharin was found, in experiments carried out in Canada, to cause cancer in mice after injection in very high doses. (In human terms, these corresponded to consumption of about 1,700 bottles of a diet carbonated beverage daily for a lifetime.) According to the Delaney Clause, saccharin should have been banned in the U.S. based on these results. In this case, however, the public outcry appears to have swayed the legislators' minds towards a more realistic approach, taking into consideration the long history of apparent safe use under actual conditions of human consumption. In other cases (cyclamate sweeteners in 1969, Red No. 2 in 1975) the Delaney clause was applied and these substances were banned as a result of similar tests. In yet another case concerning the use of nitrite in meat products (Chapter 7) the banning of the preservative would introduce risks to human safety which appear to be greater than risks of its continued use. The ongoing "nitrite dilemma," as well as the saccharin case, may also illustrate the two major aspects of the modern regulatory process: benefits of any substance allowed in foods must be clearly established and must outweigh any possible risks that could result from its use. This *risk-benefit* concept of food safety, and the testing procedures required for approval of food additives, are both essential components for the regulatory decisions.

9.5 PROCEDURES FOR TESTING FOOD ADDITIVES

Before a chemical compound is approved by government regulatory agencies as a food additive, it must be thoroughly evaluated. The general philosophy governing the regulatory process is in agreement with the international position of the UNO bodies dealing with foods (FAO, WHO, Codex Alimentarius). Food additives must be safe for continued use, must not be used for deceptive purposes, and must lead to improvements of nutritive value, food availability, or some aspects of quality that will benefit the consumer.

The current testing procedures for food additives are stringent and complicated. A summary of the data required in the regulatory process is given in Table 9-5. The tests include animal studies of short-term acute toxicity, long-term chronic toxicity, fertility, reproduction, and other biochemical and physiological tests as specified by the regulatory process. At least two different species of animals must be used; typical laboratory animals such as rats, mice, rabbits, cats or minia-

Table 9-5 Data Required[a] for Approval of a Food Additive

CATEGORY OF DATA	EXAMPLES
Description and properties of proposed substance	Chemical name, method of manufacture, chemical and physical properties, composition and specifications, proposed labelling
Proposed use	Amount, purpose, and directions for use, maximum limit for residues
Regulatory requirements	Acceptable method of analysis
Efficacy data	Proof of the intended effect
Safety data	Detailed reports of animal tests (lethality dose, subacute dose, chronic effect dose, reproductive tests with several generations)

[a] Based on requirements of the Canadian Food and Drug Act and Regulations, HPB, Ottawa.

ture pigs are most often the subjects of these experiments. The animals are sacrificed at the end of the tests and their tissues and organs are inspected for possible abnormalities.

One purpose of these studies is to ascertain a dosage level that causes no short-term or long-term effect in any of the animals. Once this level is found, a maximum suggested intake level for humans (the *Acceptable Daily Intake*) is established using a 10^2 safety factor. The calculation is straightforward. If, for example, a substance is found to have no effect in animals when ingested at 200 mg/kg body weight, the Acceptable Daily Intake for humans would be approved as 2 mg/kg body weight. According to U.S. (but not Canadian) food laws, a substance, found during large dose studies with animals to be a possible cause of malignancies, cannot be approved for food use at any dosage level.

Although the present regulatory system protects the consumer against any possible health hazards, it is cumbersome, costly, and as a result, discourages any innovative technological progress in the area of food additives. A major drawback of the present system is its vagueness regarding the evaluation of risks of high-dose levels as compared to the benefits that would result from usage of an additive at the Acceptable Daily Intake Level. Explicit considerations of the risk-benefit concept in the food additive regulatory process have been advocated by numerous opponents of the present system. Table 9-6 itemizes some of the risks and benefits that would be taken into consideration. Under this concept, all substances used as foods—"natural" foods, GRAS substances, as well as any existing and new additives—would be subject to the same food safety evaluation process. This seems reasonable since many existing foods (such as potatoes, alcoholic or caffeine-containing

Table 9-6 Some Risks and Benefits of Food Additives' Use

RISKS (ALL HEALTH-RELATED)	BENEFITS
Toxicity above threshold level	Health-related benefits
Cancerogenicity	Increased food supply
Genetically transferable changes	Reduced cost of food
Food-borne illness	Nutritional and preventative benefits
Food poisoning	Convenience
Nutritional deficiencies in newborn	Greater acceptability
Nutritional deficiency disease	

beverages cited above) would have to be evaluated for safety and an Acceptable Daily Intake level would have to be established for their biologically active component (alcohol, caffeine, solanine). On the contrary, new and more desirable additives could be approved if anomalous results from the unreasonably large testing doses used today were not the overriding criterion as required in U.S. law by the Delaney Clause. Considerable benefits could accrue to consumers regarding food availability, shelf-life, or quality.

9.6 CHEMICAL PRESERVATIVES

Antimicrobial additives used for extending shelf-life of foods are comprised of relatively few compounds suitable for retarding microbial growth, especially in combination with other preservation techniques such as refrigeration, pH adjustment or modification of the storage atmosphere (Chapters 10 and 11).

Table 9-7 lists some of the common preservative substances approved in North America for industrial food processing and also used

Table 9-7 Preservative Substances Used As Food Additives

COMPOUND	EFFECTIVENESS	EXAMPLE OF USE
Benzoic acid, benzoates	All microorganisms	Jams and jellies
Sorbic acid, sorbates	Molds and yeast	Cheese spread
Propionic acid, propionates	Molds	Bread
Sodium nitrite	Bacteria (esp. *C. botulinum*)	Prepared meats
Pimaricin	Molds	Surface application on cheese

(knowingly or unknowingly) in home preparation of preserved foods for a long time. As an example, benzoic acid and its derivatives, *benzoates*, have been used for many generations in home-made marmalades. The benzoic acid is a natural compound present in many berries (cranberries contain up to 5% of benzoic acid). Propionic acid is another example of a naturally occurring chemical with antimicrobial properties. It is produced during Swiss cheese fermentation by *Propionibacterium shermanii*, one of the microorganisms included in the starter culture, and it is a necessary precursor for the development of the characteristic holes. The sodium or calcium salts of propionic acid are suitable as effective antimolding preservatives in bread. The availability of propionates for use in bakery products is particularly advantageous as propionates are ineffective against yeasts, thus not interfering in the dough fermentation process. Similarly, sorbic acid and *sorbates* have a limited spectrum of effectiveness predominantly against molds and yeasts. There is circumstantial evidence that some lactic acid bacteria used in fermented dairy foods may produce small amounts of antibacterial compounds assisting in the preservation of these foods.

The use of antimicrobial preservatives in human food is based on the same principle of selective biological activity as employed in medical drugs. In both cases, a certain chemical is effective against one type of biological life (microbes) while being harmless to human or other types of life. The global scientific explanation of this selective effectiveness is still lacking, since the exact model of action of antimicrobial substances in a bacterial cell is not well understood in many cases. Several hypothetical suggestions include the interference of chemical preservatives with microbial cell membranes, with the specific genetic metabolism of the microorganisms, or with the enzymatic activity of bacteria. It is likely that different types of antimicrobial substances will affect microbial life differently. As an example, the effectiveness of penicillin as a drug has been shown to be due to inhibition of cell wall synthesis, thus interfering with the growth and multiplication of microbial cells.

The use of chemical additives as food preservatives is strictly regulated. However, the views of regulatory agencies in various countries are often contradictory. Some preservatives used in Europe are not allowed in North America—the best examples are antibiotics like nisin (used in cheese) or tetracycline (used on fresh fish). No antibiotics are permitted as preservatives in Canada or the U.S. within any food. The only exception is pimaricin, recently approved for use on the surface of packaged cheese.

On the contrary, some compounds used in North America such as sodium nitrite (Chapter 7) have been banned in some European countries. In most cases, there are strict upper limits for each chemical additive specified by food legislations of the various countries. All these precautions, together with long history of safe use of the common preservatives, should be reassuring to consumers who are the main beneficiaries of this often maligned food preservation technology.

9.7 AGRICULTURAL CHEMICALS AND OTHER FOOD CONTAMINANTS

When consumers or popular press criticize the chemicals in our foods, they tend to include not only the legally defined food additives or food ingredients, but also residues of agricultural chemicals like pesticides, herbicides, or veterinary drugs. The use of chemicals in agricultural practice, necessary for large-scale production of a raw food material, is controlled as tightly as the use of food additives in foods. Moreover, maximum tolerances are established for the presence of residual amounts of these chemicals in the raw material at its harvest, to insure that no health hazard will result from a possible carryover into the food chain. A safe Theoretical Daily Intake level for humans is calculated for each of these contaminants from animal toxicity studies, and routine examination of raw food materials is carried out by laboratories of regulatory agencies to enforce compliance. Some of the more commonly used agriculture chemicals are listed in Table 9–8 together with their primary functions, to illustrate the importance of these modern agricultural tools for sustained and adequate food production.

As with any human activity, there may be occasional accidents involving spillage of toxic agricultural chemicals resulting in unavoid-

Table 9–8 Types of Agricultural Chemicals Used in Production of Raw Food Materials

CATEGORY	REASONS FOR USE
Pesticides	Protection of crop vegetations against damage by insects (caterpillars, beetles, locusts)
Herbicides	Control of weeds competing with agricultural crops for the same soil nutrients
Fungicides	Prevention of molding in crop vegetations and harvested crops
Fumigants	Protection of grains and spices against infestation with worms, larvae and other insect pests
Antibiotics	Feed ingredients for prevention of disease epidemics among agricultural animals, especially poultry

able contamination of a food. Most of these cases do not result in a human health hazard since laboratory control mechanisms and cooperation of the parties involved, motivated by self-interest and possible economic consequences, are an adequate safeguard for consumer safety.

9.8 LABELING REQUIREMENTS FOR ADDITIVES AND INGREDIENTS

Labeling of processed foods in the U.S. and Canada is regulated somewhat differently even though the principles are similar. Apart from information concerning the food name, the quantity, the manufacturer, recommended storage conditions, and information regarding shelf life under proper storage conditions, a key requirement is an ingredient list giving all compounds used in the food manufacturing process in descending order of their quantity. Only certain specific food products manufactured according to standards of identity; alcoholic beverages; foods served in restaurants; and some other food products specified by the labeling laws may be exempt from the requirement for ingredient listing. The ingredients and additives used are identified by their common names, which, in many cases, are clearly understood by the consumer. In other cases (including food additives, vitamins and other ingredients), the only name available for the listing is a chemical name which may not be familiar to the general public. Since processed foods may contain numerous natural or synthetic ingredients, the required ingredient list may indeed sound like a page from a pharmaceutical lexicon. If traditional food items (meats, fruits, milk, eggs) would be required to carry labels identifying their main chemical components, they too would be a cause of anxiety for most consumers. Table 9–9 lists several common foods and gives their chemical composition using the scientific names invented by chemists, as well as common names as they exist. Many compounds found in these common foods are the same as the compounds listed as ingredients in processed foods. However, since they were added by Mother Nature rather than by the food processor, they do not have to be listed.

The required listing of all ingredients is valuable for those consumers who may wish to avoid certain substances (natural or manmade) due to health, religious or other reasons. Unfortunately, the requirement contributed to the mistrust between the industry and the public, and a major educational task is needed to increase the consumers' knowledge so that the "benefits" of the labeling regulations can outweigh the present "risks" of ignorance.

Table 9-9 Major Components of Several Common Food Items

FOOD ITEM	SELECTED CONSTITUENT CHEMICALS	
	Scientific Name	Common Name
Cow's Milk	Casein	Milk protein
	β-lactoglobulin	Whey protein
	Immunoglobulin	—
	4-O-β-D galactopyranosyl-D-glucopyranose	Lactose (milk sugar)
	Riboflavin	Vitamin B-2
	Calcium	—
	Phosphorus	—
Fresh Apple	Sacharose (sucrose)	Table sugar
	Fructose	Fruit sugar
	Glucose	Dextrose
	Acetic acid	Vinegar
	1-4-β-2-D polyglucopyranose	Cellulose
	Polygalacturonic acid	Pectin
	Ascorbic acid	Vitamin C
	Carotene	Vitamin A
	Phosphorus	—
	Acetaldehyde	—
Roasted Coffee	Caffeine	—
	Ethanol	Alcohol
	Methanol	—
	Diacetyl	—
	Methylfuran	—
	Isoprene	—
	Propionaldehyde	—

In addition to the ingredient list, a second list of specified nutrients may be carried by some foods under the present U.S. *nutrition labeling* guidelines. In the U.S., foods for which a nutritional claim is made must show a specific content of certain nutrients, while nutrition labeling of other foods is voluntary. In Canada, a nutrition labeling proposal similar to that of the current U.S. practice is under consideration. Table 9-10 gives an example of a nutrient list required by the nutrition labeling system in the U.S.

9.9 EFFECTS OF FOOD ADDITIVES IN HUMAN NUTRITION

As a direct contributor to specific nutrient needs of humans, the technological additives are unimportant. Even when the term "food additive" is interpreted in its broadest, non-legalistic sense (thus in-

Table 9-10 Nutrient Data Included in the U.S. Nutrition Labeling Scheme

NUTRIENTS WHICH HAVE TO BE LISTED[a]	OPTIONAL NUTRIENTS WHICH MAY BE LISTED[a]
Protein (g)	Vitamin D (IU)
Vitamin A (IU)	Vitamin E (IU)
Vitamin C (ascorbic acid) (mg)	Vitamin B_6 (mg)
Thiamin (vitamin B_1) (mg)	Folic acid (mg)
Riboflavin (vitamin B_2) (mg)	Pantothenic acid (mg)
Niacin (mg)	Phosphorus (g)
Calcium (g)	Iodine (μg)
Iron (mg)	Magnesium (mg)
	Zinc (mg)
	Copper (mg)
	Biotin (mg)

[a] If nutrient labelling must be provided, i.e. when a nutritional claim is made for the food item.

cluding many traditional or new food ingredients such as salts, sugars, spices, amino acids or vitamins), the total amount of food additives used in the U.S. is less than 1 percent of the annual food consumption. Almost 95 percent of this amount is attributable to salt, sugar, and other nutritive sweeteners.

Food additives are approved for food use based on their "clean bill of health." To our best knowledge today, food additives are not deleterious to human well-being through any teratogenic, toxic or antinutritional effects. The safety record of the use of additives is enviable; there are no recorded fatalities attributable to food additives as compared to other causes of deaths in the U.S. (Table 5 in Appendix II). Natural foods may be much more dangerous to human health than approved additives. Every year, fatalities and cases of food-related illness are recorded as a result of eating poisonous mushrooms, drinking "home-brews" containing methyl alcohol, or consuming other natural foods with toxic substances present in their chemical make-up or introduced by microbial growth (Chapter 2). Many common foods contain minute amounts of naturally occurring substances that could result in toxic effects if ingested in large doses. Examples may include acetone in strawberries, acetaldehyde in yogurt, solanine in potatoes or alcohol in beer. Below the sensitivity level of a human organism to a given toxin, however, the self-cleansing defense mechanisms of our bodies render these toxins harmless. On the contrary, excessive amounts of many nutrients or other naturally occurring compounds can become dangerous to human health including table salt, vitamin A or even water (all of these were diagnosed as causes of deaths in special circumstances).

A major contribution of food additives to improved human nutrition is through their indirect effect. Increased shelf-life ascertained by a chemical preservative results in reduced amount of spoilage and thus higher quantity of available nutrients. Improved functionality of an ingredient may mean a higher yield from a given raw material, cheaper food, or better quality. Even an improved look of food through the use of the much maligned food colors may result in greater acceptability and higher consumption. Let's not forget that nutrients must be first eaten to be effective as nutrients.

A possibility that food additives, especially food colors, may be related to hyperactivity in children was proposed in 1973 by D. B. Feingold, a California allergist. In an attempt to prove or disprove Dr. Feingold's theory, millions of research dollars have been spent with little concrete result to date. A major difficulty encountered in this problem is the subjective nature of the response as measured by concerned parents, teachers or other observers. Feingold himself admitted at one of the many conferences convened to deal with the subject that direct scientific evidence for his hypothesis is lacking. The current available evidence indicates that elimination of artificial colors and flavors from diets of hyperactive children has a positive effect in only a very small fraction of cases. The effects are likely those of a "placebo," or psychological effects related to the behavioral benefits of increased attention paid to children by their parents and doctors. In rare instances, sensitivity to a specific additive may be encountered just as other people are sensitive to milk protein, peanuts, and other natural food components.

The use of food additives has resulted in substantial improvements in food quality and its consistency. New foods are being developed as a direct result of rapid advances in food science research and innovative use of many traditional and new ingredients and additives. A whole new class of foods—the *intermediate moisture foods* or IMF—has emerged as a result of transforming scientific knowledge in areas of microbial growth and water activity (a_w, see Chapter 2) to industrial practice through judicious use of food additives. These foods were developed as a response to the needs of the U.S. space exploration program. The IMF concept has evolved from scientific understanding of the traditional preservation principles of pickling, salting or sugaring. One of the oldest IMF products is in fact pemmican, the traditional food of American Indians made by mixing meats, cereals, berries and other ingredients and removing some of the available water by drying. This final product had still enough moisture to provide the required chewability, but its a_w was low enough to prevent microbial spoilage. Modern IMF products (breakfast bars, space foods, some pet food products) are produced by careful ingredient calculations (Chapter 3) and

Table 9-11 Some Substances Approved for Food Use That May be Suitable as Humectants

SIMPLE SUGARS	SALTS	OTHER
Sucrose	Sodium chloride	Glycerol
Fructose	Potassium chloride	Sorbitol
Glucose		Propylene glycol
Lactose		Polyethylene glycol

incorporating certain materials such as sugars, salts, glycerol, propylene glycol and other approved food additives suitable for water activity control. A partial list of these substances commonly known as *humectants* is given in Table 9-11.

The general population needs much more information regarding the industrial use of food additives. The risks and benefits of the food additive and food labeling legislation should be addressed by the regulatory bodies. A recent survey of Canadian consumers (Health Protection Branch, Ottawa, publication No. 48, 1980) has shown the frightening level of consumer ignorance when it comes to the use of additives. As shown in Table 9-12, about half of the nearly 25,000 consumers surveyed thought (wrongly) that food additives are used in fluid milk, frozen peas or fresh meat. Although some of these opinions may have resulted from the differences between the legal definition of a food additive and the popular perception, the survey did indicate that a major educational effort is needed to convince consumers that food additives are used for the benefit of us all.

Table 9-12 Results of a Consumers' Opinion Survey Regarding Use of Food Additives in Canada[a]

QUESTION: DO THE FOLLOWING FOODS CONTAIN ADDITIVES?	ANSWER: YES (% OF ALL ANSWERS)[b]
Fresh meat	52
Soda crackers	72
Frozen peas	48
Fluid milk (2% b.f.)	56
Processed cheese	79
Canned orange juice	64

[a] Source: Publication No. 48, Health Protection Branch, Health and Welfare Canada, Ottawa, 1980.
[b] 25,000 consumers surveyed.

218 Food Additives and Food Ingredients

9.10 CONTROL QUESTIONS

1. Explain, in your own words, the term "functional property" or "functionality." Give some examples of functional properties important in food processing, and of ingredients used to achieve these functional properties.
2. What is the difference between food additives and food ingredients?
3. How are food additives produced? Give some examples of food additives found in "natural" foods.
4. What is the Delaney Clause? Is it a scientifically realistic piece of legislation? Why or why not?
5. Explain the idea of the "risk-benefit" approach to evaluation of food additives.
6. Is the philosophy of the use of food additives in the U.S.A. and Canada in agreement with international consensus? Why or why not? What are the three major principles that must be satisfied before a food additive may be permitted for use?
7. A proposed new additive was found to cause kidney stones in 85% of the male offspring of mice when fed at 5% of the diet of the test mice. No effect of any kind was found in feeding tests with four generations of the same mice and six generations of monkeys when the dose used was 1% of the diet. Should this additive be permitted for use in human foods? If no, give a thorough argument. If yes, calculate the acceptable daily intake (ADI) level for humans if the average amount of food eaten daily by the test mice (weighing 250 mg each) was 100 g, and that of the monkeys (weighing 2 kg each) was 800 g.
8. Explain the difference between "ingredient labelling" and "nutrition labelling." Are all U.S. and Canadian foods labelled for their nutrient content? Do you think that it is necessary to have all foods nutritionally labelled? Why or why not?

Suggested reference books—
Food Additives, Food Ingredients, Functional Properties.

Anon. 1981. *Encyclopedia of Food Chemicals*. Toronto: Food in Canada.
Cherry, J. P. 1981. *Protein Functionality in Foods*. Washington, D.C.: American Chem. Society.
Clydesdale, F. 1979. *Food Science and Nutrition*. Englewood Cliffs, New Jersey: Prentice-Hall.
Conners, C. K. 1980. *Food Additives and Hyperactive Children*. New York: Plenum Press.

Health Protection Branch. 1982. *Food Additives in Action — A Resource Guide.* Health and Welfare Canada, Ottawa: Educational Services.

Health Protection Branch. 1977. *Guide to Food Additives.* Health and Welfare Canada, Ottawa: Educational Services.

Orr, H. L. and Murray, D. B., Publ. 1498, 1977. *Eggs and Egg Products.* Agriculture Canada, Ottawa: Research Branch, publ. 1498.

Packard, V. S. 1976. *Processed Foods and the Consumer.* Minneapolis: Univ. of Minnesota Press.

Stadelman, W. J. and Cotterill, O. J. 1977. *Egg Science and Technology.* Westport, Conn.: AVI Publ. Co.

10

Processes for Food Preservation

10.1 CAUSES AND PREVENTION OF FOOD SPOILAGE

Food, as all organic matter, is subject to degradative changes caused by our environment. The major type of food spoilage is *microbial*, caused by growth of bacteria, yeasts, or molds. *Chemical* spoilage can also occur. Typical examples are development of rancidity in foods containing fat, various enzymatic reactions in fruits, vegetables, meats, or milk products, or the so called non-enzymatic browning ("Maillard") reactions occurring in the heating of foods containing protein and sugar. Water, oxygen, light and temperature play major roles in chemical spoilage. Spoilage due to *physical* causes may include mechanical damage, drying out, loss of texture, or even infestation by insects or rodents. Of the three types of spoilage, deteriorative changes due to rapid growth of microorganisms normally proceed at much faster rates than chemical or physical changes. Thus, the primary objective of food preservation techniques is to stop or retard microbial spoilage, while minimizing chances for enzymatic deterioration and physical changes.

Proper storage conditions, especially low temperature, darkness, insect-proof containers or controlled atmosphere storage (CA, see Chapter 4), offer limited protection against spoilage of fresh agricultural products for a relatively short time. Special processing techniques must be used for long-term preservation of the often large oversupply of food produced at certain times of the year. Freezing, canning, and drying are the three principal food preservation techniques used today. Many other food manufacturing processes discussed in the preceding chapters also include an inherent preservation aspect. The baking of bread, manufacture of ice cream, production of fruit jams, fermentation of yogurt, smoking of sausage and many other processes result in foods with prolonged shelf-life. These techniques, however, are more properly classified as manufacturing since their principal goal is creation of a new food product. Freezing, drying, and canning are used to protect all foods (raw agricultural produce as well as manufactured food items) from microbial, chemical or physical spoilage for many months. The basic preservation principles of several food processing techniques are mentioned in Table 10–1.

To prevent microbial spoilage we must stop *microbial growth*. Mere presence of live microorganisms will not spoil food as long as the bacteria, yeasts, or molds are unable to grow and multiply. There are many kinds of spoilage microorganisms, and each has its specific growth requirements for proper temperature, presence or absence of oxygen, availability of specific nutrients, proper acidity (pH) of the environment, and availability of water, as discussed in detail in Chapter 2.

The purpose of the three major food preservation techniques is either to alter the environmental conditions such that microorganisms will not be able to grow (freezing, drying) or to destroy the viability of the microorganisms present (canning). None of the three techniques ac-

Table 10–1 Basic Preservation Principles of Main Food Processing Techniques

PROCESS	PRINCIPLE OF PRESERVATION
Drying	Removal of water necessary for growth of microorganisms
Freezing	Solid water (ice) in food unavailable to microorganisms; low temperature of storage inhibits microbial growth and many enzymatic reactions
Refrigeration	Microbial growth stopped or slowed down (psychrotrophs) by low temperature
Heat sterilization	Inactivation of all viable forms of microbial life in the food; hermetic packaging prevents recontamination
Other heating processes (baking, pasteurization, smoking, frying)	Inactivation of many viable microorganisms
Fermentation	Lowering of pH by acid-producing microorganisms used as culture
Pickling, salting and sugaring	Lowering of pH and/or a_w below microbial growth tolerance limits by added acidulants, salts or sugars

tually eliminates the microorganisms from food; improperly dried, frozen or canned foods will show signs of microbial spoilage readily.

The major causes of chemical spoilage are food enzymes, various deleterious reactions involving water (e.g. hydrolytic rancidity of milk fat), or many oxidative reactions, principally between food fats and atmospheric oxygen. The preservation principles inherent in the processing techniques used to prevent the microbial growth are often unsatisfactory with respect to chemical spoilage—and, in many cases, will enhance it without proper control. Thermal destruction of enzymes before freezing, elimination of oxygen from canned foods, or use of chemical additives preventing browning and lipid oxidation in some dried foods are just three examples of how chemical spoilage can be minimized after the danger of microbial spoilage has been eliminated.

None of the principal preservation techniques is perfect. All have their advantages and drawbacks in affecting certain nutrients, creating environments suitable for undesirable reactions, or causing major changes in food texture and other quality aspects. The selection of the most appropriate technique for a given product is influenced by the severity of these undesirable side effects, as well as by process engineering economy and market acceptance of the final product. In many cases there may be no significant advantage for any of the three processes (supermarket shelves offer dry, as well as canned and frozen peas), while many examples can be found where one technique is clearly superior. Because of the complexity of problems encountered in long-term preservation, many alternative engineering designs based on a common technical principle are available for a given process. The various technological alternatives of each process may have a profound effect on the final product quality. In this text only the general principles and their industrial applications will be briefly explained for the main preservation processes as they relate to consumer attributes of processed foods.

10.2 FOOD DEHYDRATION

Natural drying of food by the sun has been the oldest preservation technique used by mankind. The broader term *dehydration* encompasses all modern drying techniques including the use of hot air in various types of air driers, use of hot surfaces as in roller drying, or sublimation of ice from frozen food (freeze-drying). Table 10–2 lists and briefly describes the principles of major drying operations.

The effectiveness of drying as a preservation technique is due to the removal of water, present in all perishable foods and required by all microorganisms for their growth. Since water is readily available for

Table 10-2 Technological Principles of Drying Operations

OPERATION	PRINCIPLE
Sun drying	Radiant heat from the sun evaporates water from food to surrounding air
Hot air drying	Stream of hot air in closed environment supplies that heat by convection and carries away the evaporated water
Spray drying	Special case of hot air drying; fluid foods sprayed into the stream of hot air; dry powder separated mechanically
Fluidized bed drying	Another special case of hot air drying; beds of particulate foods airlifted by hot air for better heat and mass transfer
Vacuum drying	Removal of water vapor by creation of vacuum; heat must be supplied by conduction or radiation (no air used!)
Roller ("drum") drying	Liquid or pasty foods applied on surface of hot stainless steel rolls; dry material scraped off the rotating rolls at the end of a revolution
Freeze-drying	Water removed by sublimation from frozen foods under high vacuum; heat must be supplied by conduction or radiation through the dry food layer

most consumers, dried foods may be reconstituted to their original composition quite easily at the point of final use. Water removal from dried foods is advantageous for the substantial reduction of their volume and weight since water is often the major food component. This is important for the economics of transportation and prolonged storage. Drying is not overly damaging to microorganisms. After reconstitution dried foods will be as perishable as the fresh product.

While the principle of food dehydration is quite simple, its technological execution, including the control of concomitant food quality defects, is not easy. The major deteriorative effects of drying are physicochemical, resulting from the need for sufficient heat to vaporize and remove the liquid water from food. (In engineering terminology, we speak of simultaneous heat and mass transfer.) The required *heat of vaporization* is quite large, about 540 Kcal or 2,200 kJ/kg H_2O (see Table 1-7 for definitions of these units), and must be supplied to water molecules in food to make them volatile, thus producing vapors that can be easily removed. Unfortunately, many food flavor compounds are also volatile—even more than water—and these will be removed together with the water in the process. The result may be a much less flavorful final product. For many foods where flavor is the primary quality determinant (soluble coffee, onion powder), the choice of suitable drying processes may be severely limited.

Due to water removal from the structure of the food being dried, the void spaces left behind may collapse and the food particle will shrink. The resulting dry food may be shriveled, compact, and often impossible to rehydrate. This defect is minimized in vacuum drying when the atmospheric pressure does not exert its effect; providing heat for efficient vacuum drying, though, is difficult without air.

In most food dehydration processes, the supply of necessary heat is accomplished by hot air. In air drying, all water vaporization theoretically occurs on the food's surface, thus protecting the food interior from heat damage. In practical drying operations, this may not be the case and severe heat effects may be encountered. These include browning, heated (scorched) flavor, denaturation of proteins and other effects resulting in loss of solubility upon rehydration, textural changes, and some nutrient losses in heat labile vitamins. Greater vitamin losses normally occur, however, during storage of dried foods.

Various engineering designs of food dehydration operations (Table 10-2) have tried to overcome some of these defects with a varying degree of success, often at substantial cost. One of the most sophisticated techniques, *freeze-drying*, is based on removal of water from previously frozen food by sublimation, when the water molecules are vaporized directly from the solid state. This results in relatively little heat damage, little loss of volatile flavor substances, and good rehydratability since the frozen food material is immobilized during the drying process and thus cannot shrink. The process is very expensive due to its technical nature, the complexity of the equipment, and the slow water removal. At the present time the freeze-dried food market is limited mainly to items manufactured for special purposes (mountaineering, army rations), foods of a high prestige such as instant coffee that is expensive even without freeze-drying, or certain ingredients for convenience foods (chicken meat or mushrooms for dry soup mixes) which require quick rehydration and/or superior flavor. Freeze drying may not be the preferred technique in situations where preservation of texture is important (some vegetables, delicate fruits) since the foods must undergo the freezing step first and this affects the texture (Section 10.3).

The most widely used drying technique suitable for liquid foods is *spray drying* (Figure 10-1) involving spraying small droplets of fluids and slurries (milk, eggs, coffee extract, cheese slurry) into a continuous stream of hot air. The air is used both as a heat supply medium and as a transport medium for the water vapor evolving from the fast-drying droplets.

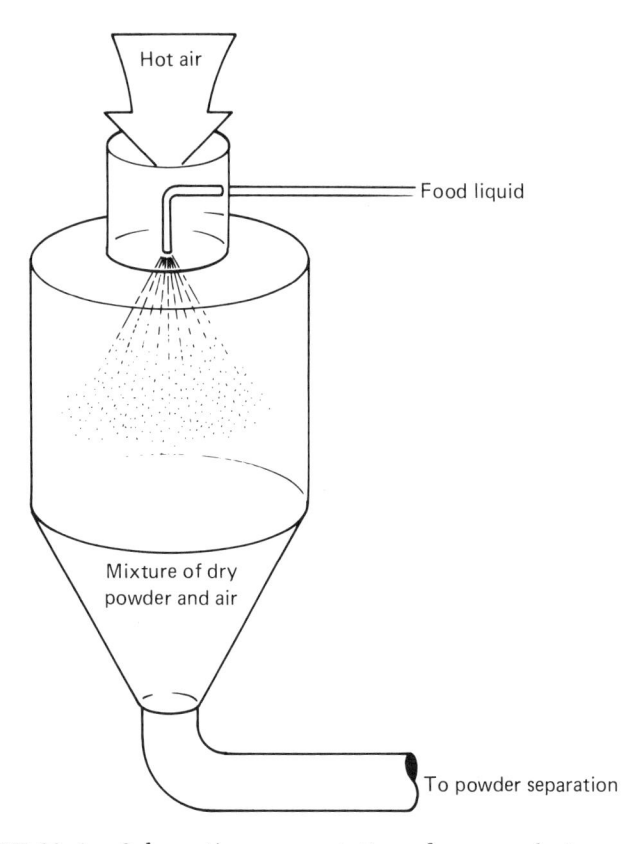

FIGURE 10-1. Schematic representation of a spray-drying operation.

Drying occurs in a matter of seconds and this results in little if any heat damage to the product. The rapid nature of the drying process is due to the very large surface area of droplets produced in the spray mechanism. As all of the heat and mass transfer occurs through the droplet surface, the larger the total droplet surface area, the more efficient the drying. A relationship between droplet size and the resulting total surface area created by the spray is shown in Table 10-3. The separation of dried powder particles from the air is accomplished by various mechanical means. The design and operation of a spray drier is a complex engineering task; proper adjustment of the dry particle size is one of the critical control variables affecting possible losses of powder in the exit air, as well as rehydration qualities of the dry powder. Very small particles are easily entrained in the exhaust and are more difficult to reconstitute in water (clumping of regular dried skim milk upon rehydration is a well known irritant of many housewives). On the contrary, too large particles may not be properly dried,

Table 10-3 Surface Area of Droplets Produced in Spray Drying From 1,000 L of a Liquid

DROPLET DIAMETER (μm)	NUMBER OF DROPLETS	TOTAL SURFACE AREA (m²)
1,000	2×10^9	6×10^3
100	2×10^{12}	6×10^4
1	2×10^{18}	6×10^6

thus impairing the shelf-life of the dry powder. A special technique called agglomeration is frequently used to create large clumps from small particles of dry milk, chocolate or coffee, which will rehydrate very rapidly (thus "instantized products"). The technique consists of partially moistening the surface of previously dried particles in an atmosphere of wet steam. When the sticky particles are brought to collide in the turbulent atmosphere of the instantizing tunnel (Figure 10-2), they will produce an agglomerate with void spaces which will allow the water to readily penetrate throughout the structure. The difference between regular and instantized milk powder is shown in Figure 10-3.

FIGURE 10-2. An agglomeration tunnel used for production of instantized dry milk. (Photo courtesy of Niro Atomizer, Denmark.)

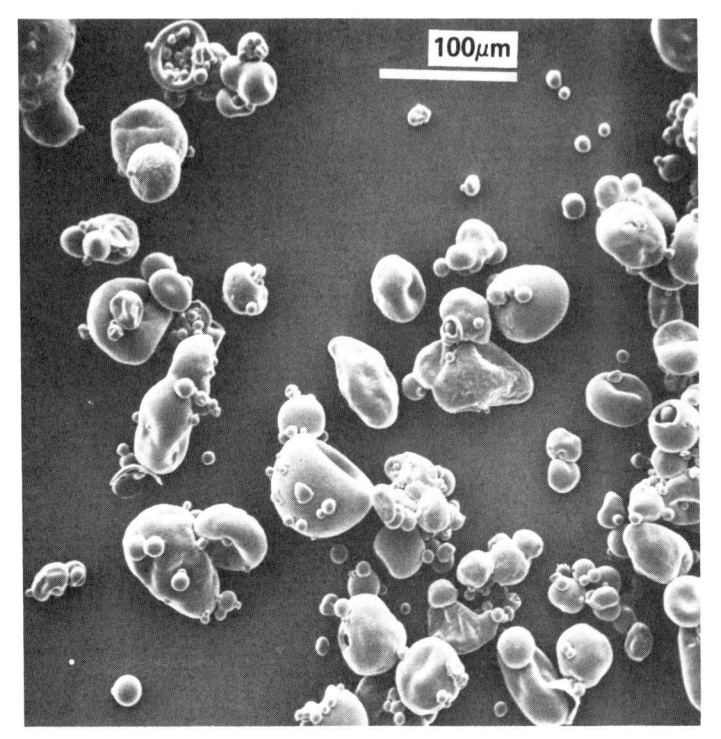

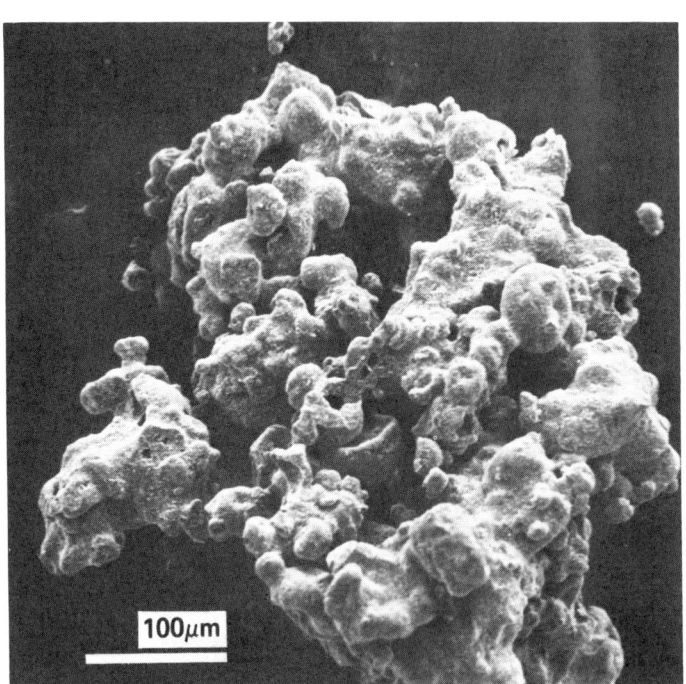

FIGURE 10–3. Microscopic illustration of differences between (A) regular and (B) instantized milk powder. (Photo courtesy of Dr. M. Kaláb, Food Research Institute, Ottawa.)

Packaging of all dried foods is an important task especially for those foods that contain fat or are hygroscopic (will reabsorb moisture from air if unprotected). Lipid oxidation in dehydrated foods is the most frequent cause of their spoilage. To minimize this problem, special packaging is used for many dehydrated foods with high fat content. Frequently used precautions include filling the package with nitrogen (Figure 10–4), creating a vacuum in the package, or enclosing a separate packet of a chemical additive which will preferentially react with the available oxygen thus protecting the lipid from being oxidized. Most dry foods including soups, dough mixes, pasta products, various beverage powders, raisins, and many industrial food ingredients (onion powder, protein concentrates) combine the preservation aspects with convenience for transportation, food preparation and further use.

FIGURE 10–4. Filling machinery for packaging dry milk products under nitrogen gas.

10.3 FOOD FREEZING

The development of the modern food freezing industry in North America had its beginnings in the 1920s and is connected with the names of Clarence Birdseye in the USA and Bill Heeney in Canada, two of the true pioneers of the food freezing process. The rapid expansion of the technology occurred hand-in-hand with engineering developments in the field of refrigeration equipment and the widespread use of household refrigerators. Food preservation by freezing requires both expensive processing and expensive storage facilities; however the quality deterioration due to freezing is generally less severe than in food dehydration or thermal processing.

In principle, the preservation aspect of food freezing is the same as in drying: liquid water is made unavailable for growth of microorganisms, this time by immobilization as ice. The low temperature at which frozen food is held is also an effective deterrent of microbial growth. The principal difference between refrigerated and frozen food storage is in the physical state of the food. The frozen food is held below its freezing point, when most of the water contained in the food has turned into pure ice crystals. In refrigeration, the water in the food is still in the liquid state and thus available for psychrophilic spoilage microorganisms (Chapter 2) that can grow at refrigerated temperatures. Freezing is not effective in destroying microorganisms, although some reduction in viable microbial populations is usually achieved in commercial food freezing. After thawing, the previously frozen foods must be handled in the same manner as if they were never frozen.

The main damaging effect of freezing on food quality concerns food texture. This is due to the formation of individual ice crystals within the food. The dry matter components (Chapter 1) with the possible exception of fat usually undergo no phase change upon food freezing. The mere appearance of sharp ice crystals in the cellular structure of a plant tissue or an animal muscle is not considered to be the main damaging effect. It is well known, however, that when water freezes, it expands by 9 percent in volume, and the size of the individual crystals related to the freezing rate (rapid *vs.* slow freezing) may have a profound effect on the textural changes. In general, the faster the freezing, the smaller the ice crystals will be and the less the structural damage. The widespread practice of keeping fresh foods (often bought as discounted specials) in a household freezer represents one of the slowest freezing methods available. The resulting textural defects (accompanied by loss of nutrients in the juice expelled after thawing the damaged texture) may make this practice philosophically questionable when compared with the efficient industrial freezing process.

The mechanisms of ice formation and the resulting textural breakdown are related to the phenomenon of *freezing point depression* caused by dissolved substances. It is well known from everyday life that when salt, sugar, glycol, alcohol or other simple solutes are added to water, the solution will freeze at a lower temperature than pure water. This freezing point depression ($\triangle$T) is proportional to the concentration of the solute molecules in water (M), as expressed for an idealized aqueous solution by the following freezing point depression formula (Equation 10-1):

$$\triangle T = -1.86 \times M \{°C\} \qquad \text{(Equation 10-1)}$$

As water freezes at 0°C, the freezing point depression $\triangle$T (which is the difference between the freezing point of water and the freezing point of a solution in which the dissolved substance is present in molal concentration M) gives directly the new freezing point. For ideal solutions containing *1 mole* of any solute the freezing point will be −1.86°C. In real solutions, since no ideal solutions exist, deviations are encountered due to dissociation of molecules and other physicochemical phenomena.

Although foods are multicomponent systems where the constituent water contains many dissolved substances, the principles of the freezing process are the same as in the simple solutions, and can be followed in a theoretical representation called *phase diagram* as shown in Figure 10-5. When foods are placed into a freezer, their constituent water starts forming pure ice crystals when the food temperature reaches the apparent freezing point determined by the composition of the food (point A in Figure 10-5 indicating 80% water and 20% dissolved substances in this case). As the temperature of the food decreases further (ultimately reaching the temperature of the freezer), additional water freezes out as ice, thus leaving behind more and more concentrated solution of the food solutes. In Figure 10-5, at −5°C the unfrozen solution in the system contains approximately 50% water and 50% solutes; the remainder of the water is present as pure ice (naturally, the ice crystals form throughout the system; since no ice is physically removed the overall composition is still that corresponding to point A, i.e. 80% H_2O and 20% solutes). No unfrozen solution can exist below a certain point called *eutectic point*. For a pure water-sucrose solution this has been determined experimentally as −9.5°C; multicomponent food systems often have much lower eutectic points. The exact numerical values of freezing points, eutectic points and corresponding composition data are highly dependent on the type of solutes predominant in a frozen food; some apparent freezing points of common foods are given in Table 10-4.

232 Processes for Food Preservation

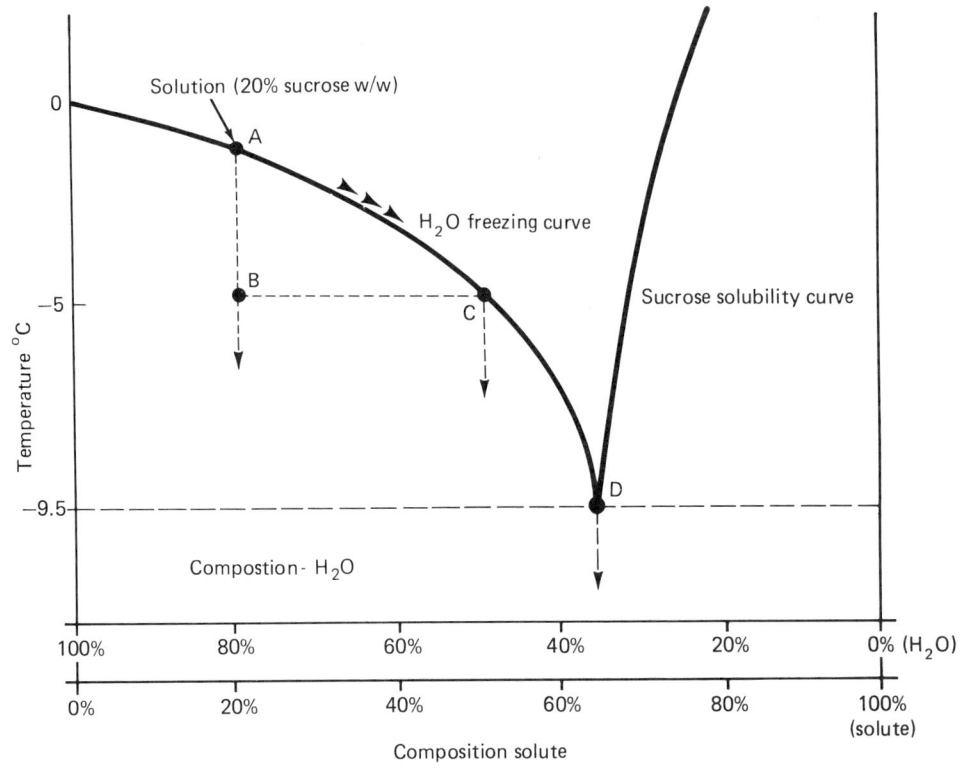

[Figure: phase diagram with temperature (°C) on y-axis from 0 to below -9.5, and composition on x-axis showing % H₂O from 100% to 0% and % solute from 0% to 100%. Points A, B, C, D labeled on curves for H₂O freezing curve and Sucrose solubility curve. Solution (20% sucrose w/w) indicated at point A.]

Notes:
A = apparent freezing point of the solution.
B = composition of the mixture (pure ice + unfrozen sucrose solution).
C = composition of the unfrozen solution.
D = eutectic temperature and composition (no liquid solution below this point).

FIGURE 10-5. A theoretical liquid-solid phase diagram for sucrose-water binary solution.

The "freeze concentration" effects explained in Figure 10-5 are important in several aspects of the food-freezing process. The texture of some foods may be improved by freezing them in sugar solutions (strawberries, raspberries) since the more concentrated solute system inside the tissues will not form damaging ice crystals so readily. On the contrary, even the most efficient freezing process will not solidify all of the water in any food unless the eutectic temperature is reached. The remaining water, still in liquid state, may be available to some microorganisms (psychrophilic spoilage) or, more importantly, to enzymes for enzymatic deteriorative reactions. Thus, all commercially frozen vegetables must be blanched before freezing. For some products

Table 10-4 Apparent Freezing Points of Some Common Foods[a]

PRODUCT	APPROXIMATE FREEZING POINT (°C)
Fresh meats, poultry and fish	−1.5
Fresh fruits (cherries, plums, peaches)	−2.0
Vegetables (beans, peas, cauliflower, etc.)	−1.0
Fluid milk	−0.5
Cheese	−2.5
Ice cream	−6.0
Eggs	−2.5
Lettuce	−0.5
Dates	−20.0
Beer	−2.2

[a]Average values compiled from various sources.

(e.g. mushrooms) the use of chemical additives is permitted since blanching would enhance the severe textural problems caused by freezing. Quite substantial color changes, development of bitterness, and other defects resulting from enzymatic activity during frozen storage are not uncommon in improperly processed frozen foods. The unfrozen concentrated solution of organic acids and salts present in living tissues may be injurious to the cellular structure of some frozen food materials thus resulting in further textural damage.

Fluctuation of frozen storage temperatures will cause melting and repeated freezing of some of the frozen water. (In Figure 10-5, composition of the unfrozen solution will change along the CAC pathway.) The thawing will eliminate some of the smallest ice crystals first. Upon repeated freezing, the water will not form another crystal, but will deposit on some of the larger crystals already available. Thus, repeated temperature fluctuations may cause growth of relatively large ice crystals; this may be observed in improperly stored ice cream.

Proper packaging of frozen food is of utmost importance for prevention of sublimation of ice from surface food layers. This phenomenon is similar to freeze drying and results in a condition called *freezer burn*, denoting the formation of dried, hard and unappealing food surface left behind by the sublimed water. The sublimation, and the subsequent deposition of ice on the freezer coils of the commercial or household cold storage equipment, is caused by the temperature gradient—and correspondingly the a_w gradient—between the food and the coldest spot in the refrigerator. This will force some of the water from the food to evaporate and migrate to the point of the lowest a_w if water vapor permeable packaging is used (Chapter 11).

Industrial freezing equipment is designed for an efficient removal of heat from food. In freezing, heat transfer proceeds in the direction which is opposite to that of drying. Still or moving cold air is most often used as the heat transfer medium to remove the latent heat of fusion resulting when the water in the food forms ice. Other freezing systems employ a highly concentrated salt solution (brine) or other liquids with naturally low freezing points such as alcohol or glycerol. Table 10-5 lists and briefly describes the main industrial freezing systems.

Table 10-5 Technological Principles of Freezing Operations

FREEZING OPERATION	FREEZING RATE	PRINCIPLE
Sharp Freezing	Slow	Food placed into still air of low temperature
Air Blast	Intermediate	Food placed into stream of circulating cold air in a tunnel
Fluidized Bed	Intermediate	Similar to fluidized bed drying, only using cold air
Contact Plate	Slow or Intermediate	Rectangular packages pressed between two plates cooled by liquid freezant
Scraped Surface	Intermediate to fast	Pasty or particulate foods frozen by freeze adhesion inside or outside of a cold cylinder surface scraped continuously by a rotating blade
Liquid Immersion	Intermediate to fast	Packaged or unpackaged foods immersed into cold liquid of a low freezing point (glycerol, aqueous solutions of sugars or salts, liquid freons)
Cryogenic Freezing	Very fast	Food sprayed with a cryogenic liquid of extremely low boiling point (liquid N_2 or CO_2)

The heart of virtually all current industrial cooling and freezing operations—as well as the basis for home refrigerators—is a mechanical refrigeration system shown in principle in Figure 10-6. The system is based on evaporation of a suitable liquid having low boiling point temperature at ambient pressure (normally *ammonia* in industrial conditions and a special non-toxic liquid *freon* in home refrigeration). As seen in Figure 10-6 (point A), the evaporation occurs at a point where the cooling effect is desired, inside coils of an evaporator. The latent heat, needed for the evaporation, is taken from the food via the heat transfer medium (the air, the brine or the glycerol) surrounding the evaporator. From the evaporator coils the refrigerant vapors are taken by a mechanical compressor (point B) and compressed to a high pressure, thereby increasing their condensation temperature to levels attainable by conventional coolants. The compressed vapors proceed into the condenser (another set of coils cooled by ambient air or city water) where their latent heat of evaporation is taken away (point C). The condensed liquid, still under high pressure, is returned to the system; upon passing through a throttling valve (point D), the original low pressure and correspondingly the low boiling point is again attained. Naturally, the mechanical refrigeration is a fully closed system with no direct contact between the coolant and the food or the environment. Since ammonia is quite toxic, the industrial systems must be constantly moni-

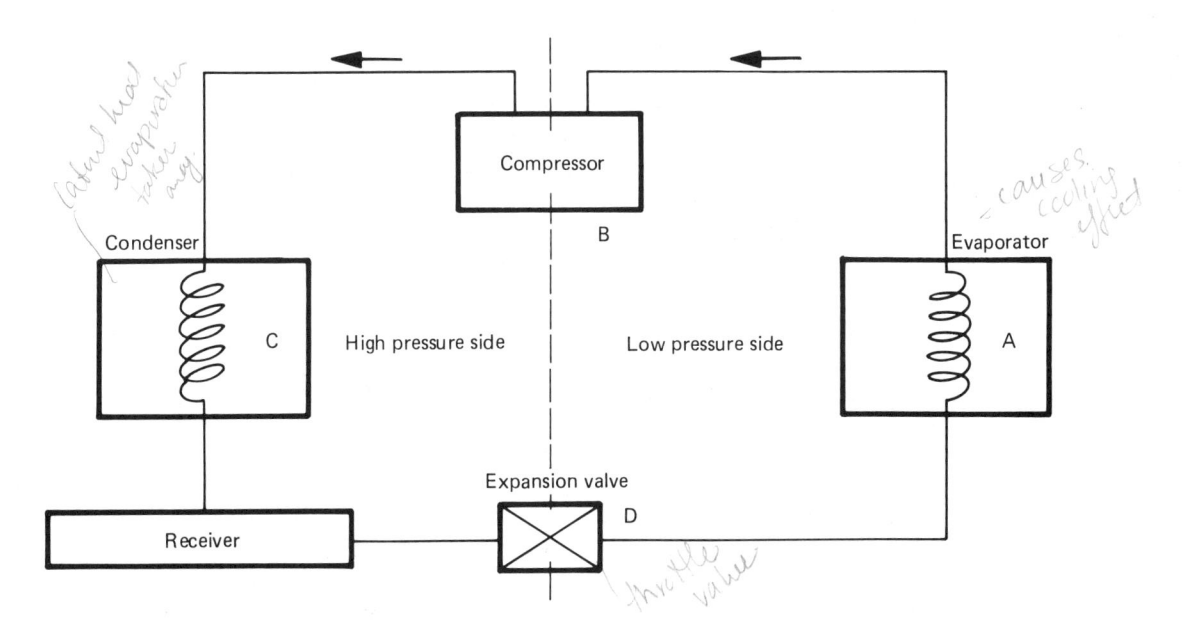

FIGURE 10-6. Principle of a mechanical refrigeration system.

tored to avoid any possible leakage or contamination of the processing plant with ammonia.

A newer *cryogenic freezing* system which is based on the direct use of liquid nitrogen (boiling point $-196°C$) of liquid CO_2 (b.p. $-120°C$) as freezants is slowly gaining industrial popularity. Its extremely rapid freezing rate produces frozen foods of good quality and it eliminates the need for the expensive mechanical refrigeration equipment. The operating costs of the cryogenic freezing, however, are generally high due to the recurring need for the non-recoverable cryogenic gas.

Freezing as a food preservation technique is not damaging to any food nutrients, while the requirements for blanching may leach out some micronutrients, especially vitamins and minerals. Further losses of vitamins may occur in prolonged storage.

Losses of over 60% vitamin C and over 40% of some of the B vitamins were recorded in literature after six to twelve months storage of fruit, vegetables and meats. Other nutrients—minerals, proteins, fats, and carbohydrates—are not significantly affected by frozen storage. On the whole, freezing appears to be one of the best, but also one of the most expensive commercially available methods for long-term food preservation.

10.4 HEAT STERILIZATION AND CANNING

Canning as a food preservation technique has its origins in the early 1800s after a French confectioner, Nicholas Appert, observed that food heated in sealed containers did not spoil. The discovery, although unexplainable by contemporary food science, found an immediate use in feeding the armies of the French general Napoleon. In the ensuing 150 years, canning became the most widespread food preservation technique which influenced the eating habits of the industrial world.

The explanation of Appert's success came more than half a century later from another Frenchman, Louis Pasteur, who identified microorganisms as the cause of spoilage in unprocessed foods. Canning makes use of the fact that bacteria are susceptible to heat. Destruction of all forms of viable microbial life by heating the food in hermetically sealed containers, thus preventing bacterial recontamination after heating, is the principle of food preservation by canning. The relatively severe heating required to kill the microorganisms is also sufficient to control enzymatic changes and other causes of chemical spoilage. Furthermore, to prevent difficulties in heating the sealed containers, and to eliminate a possibility of oxidative changes, the cans are evacuated before heating. For all these reasons, canned foods are among the most shelf-stable industrial food products and their shelf life may be virtually indefinite, as evidenced by the occasional recoveries of perfectly acceptable canned food items from shipwrecks 100 or more years old.

Since canning is based on the use of heat, it is similar in principle to pasteurization, blanching, baking, and other processes described earlier. However, the aim of canning is to achieve *commercial sterility*, or destruction of all pathogenic microorganisms as well as more heat resistant organisms that could grow inside the processed can under normal conditions of distribution and room-temperature storage. The amount of heat that the whole content of the can must receive to achieve commercial sterility is much greater than in other heated foods due to high heat resistance of many spoilage bacteria and especially their spores (Chapter 2).

The determination of the appropriate time and temperature of heating for various canning processes is a rather involved mathematical procedure, based on thermal resistance of the individual microorganisms present, and specific engineering data for heat transfer through foods and the different types of containers used. To minimize the heat damage of various heat-labile nutrients (principally vitamins), the heating conditions must not be excessive, yet adequate to kill all microbial contaminants that would result in spoilage or health hazard.

Particularly important is the need for destruction of spores of *Clostridium botulinum*. This organism grows in the absence of oxygen and above pH 4.6, environmental conditions found in many canned foods. *C. botulinum* is the only food-related microorganism that produces a deadly toxin, and *botulism*, the most dangerous and often fatal type of food poisoning, can be caused by contaminated canned food. The botulotoxin can be produced by the actively growing *C. botulinum* organisms in cans that were improperly processed, usually due to inadequate heating. Sometimes, spores of certain spoilage bacteria will also survive in—and eventually spoil—the under-processed can, thus providing a desirable safeguard. Fortunately, the toxin is heat labile and would be destroyed when canned foods are heated before eating. Botulism poisoning is a very rare phenomenon considering the enormous amount of canned foods consumed. Most of the individual cases reported in the press were caused by home-canned foods, or by products such as canned tuna or canned soup that were eaten unheated. If a can of food is suspect for any reason—e.g. bulging can ends, spoiled (frothy) appearance, unnatural smell, etc.—it should not be eaten or even tasted without adequate prior heating. "If in doubt, throw it out" is a useful general rule applicable to all canned foods.

Since the *C. botulinum* organism is not capable of producing toxin (and most spoilage organisms would not grow) below pH 4.6, many naturally acidic foods and some other products purposely acidified below pH 4.6 may be processed by less severe heating. This is important since the amount of heat needed for destruction of *C. botulinum* and the spoilage bacteria may be damaging to the canned food texture and micronutrient content. The selection of the appropriate temperature and time conditions for the canning process is one of the most important tasks of the food process engineer. The final choice will be influenced by the *type of food processed* (liquid foods have much higher heat transfer resulting in shorter heating than solid foods); its *acidity* (all foods of pH 4.6 or higher, referred to as low-acid foods, must be processed to commercial sterility which normally requires heating above 100°C in pressurized vessels); *size of can* (smaller cans may require less heating time); total *microbial load;* and the most predominant *type of spoilage microorganism* present (various foods may be populated by different organisms having very different heat resistance characteristics). Approximate times used commonly for industrial canning of several typical acid and low-acid foods are listed in Table 10-6.

Industrial canning is based on the existence of a *sanitary tin* can. Both of these terms are technical jargon. The sanitary can is the typical, cylindrical can produced by folding the can body from a flat piece of steel plated with a very thin layer of tin, and attaching—by a unique mechanical process—the bottom lid (Figure 10-7). These two operations are done in a few specialized sheet metal factories at a rate of several hundred cans/min. The empty cans are supplied to food processing plants where they are filled with food. The top lid is then attached in the same manner as the bottom lid, and the sealed cans are placed into *retorts* which can be described as industrial pressure cookers of various design. Table 10-7 summarizes and briefly describes the main types of retorts and other methods of food sterilization.

Table 10-6 Examples of Processing Conditions for Several Canned Foods[a]

	PROCESS TEMPERATURE (°C)	PROCESS TIME (MIN)
Low Acid Foods		
Green beans	121	12
Cream-style corn	118	105
Pork-and-beans	118	105
Corned beef	116	150
Acid Foods		
Tomato	100	45
Apple juice	84 (filling temp.)	–

[a]Source: A complete course in canning, by A. Lopez. The Canning Trade, Baltimore, Maryland, 1982. All data are for the No. 2 can (see Table 11-4). Adapted by author.

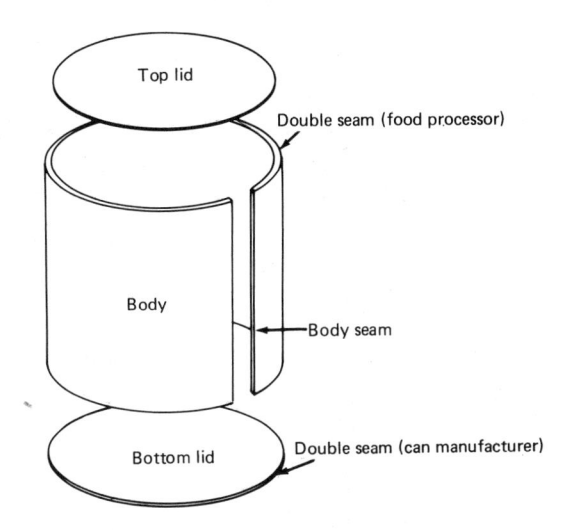

FIGURE 10-7. Diagrammatic representation of a sanitary can production.

Table 10-7 Technological Principles of Heat Sterilization and Canning of Food

EQUIPMENT OR PROCESS	PRINCIPLE
Still Retort	Vertical or horizontal pressurized vessel; sealed containers with food are heated by steam or superheated water without movement
Agitated Retort	Food containers are rotated (end-over-end or axially) inside the retort during heating)
Hydrostatic Sterilizer	Continuous retort, cans carried by conveyor through heating tower with steam, "sealed" by hydrostatic legs of water for heating or cooling
Steri-Flame Process	Cans heated rapidly by rotating above direct flame
Pouch-Pack Process	Retort sterilization of food products packaged in flexible laminated plastic containers
UHT Process	Heat sterilization of fluid foods outside the package, followed by packaging under sterile conditions
Aseptic Canning	Special Case of UHT processing, followed by aseptic filling into pre-sterilized cans

Because of the shape of the round can, much of the food in the can will be overheated in order to heat the "*cold spot*" of the can to the proper temperature for a given product. The cold spot of a can containing solid products (e.g. luncheon meat) is in its geometrical center. Cans of liquid or mixed foods (fruits or vegetables in sugar syrup or salt brine) normally have a cold spot close to the bottom due to the natural convection of the heat and the differences in density of cold and hot fluids.

One of the newly emerging alternatives to the can is a *retortable pouch* made of plastic or plastic-metal foil laminate (see Chapter 11), which is much flatter and thus its content can be heated more rapidly with corresponding reduction in the heat damage to food. The main disadvantages of the current pouch-pack systems are the much slower filling and sealing machinery, requirements for sturdy secondary packaging (Chapter 11) to protect the fragile package in transport and storage, and market acceptance.

For sterilization of homogeneous liquid foods, the ultra-high-temperature (*UHT*) processing offers an increasingly popular alternative based on heating the food outside the package which is then filled under aseptic conditions. Heating, typically at 140–144°C for a few seconds, is practically instantaneous, especially in the "*direct*" *heating systems* (Figure 10–8) where liquid food is mixed with culinary steam. This insures complete sterility without significant effects on food quality (flavor, protein denaturation, nutrient retention). The steam is subsequently removed from the heated liquid by flash evaporation in a vacuum chamber; in fact this insures almost instantaneous cooling thus further minimizing the undesirable heat effects. The use of UHT technology is inseparable from the advances in aseptic packaging and its pioneer Ruben Rausing, the Swedish founder of the Tetra-pak company whose machinery (Figure 10–9) has been the main factor in the widespread use of UHT sterilized products. The recent FDA decision permitting hydrogen peroxide as an indirect sterilant of food packaging materials (the principle used by the Tetra-pak and other aseptic packaging systems) paved the way for a rapid increase in the use of UHT technology in the United States, even though several plants have been operating in Canada for some time.

The effects of heat sterilization on vitamins and other nutrients have been extensively studied in various foods. The reported losses of vitamins in canned vegetables were highly variable, depending on the vegetable, the type of process, and other factors discussed above. In general, the losses may be quite severe—occasionally in excess of 80 percent for the more heat sensitive vitamins (Chapter 1); in other cases, only minimal vitamin losses were reported. Losses of water-soluble vitamins and minerals may also occur in the blanching process pre-

FIGURE 10–8. Ultra-high-temperature sterilization of milk by the direct system (live steam mixed with milk). (Photo courtesy of APV Crepaco Canada.)

ceding the canning operation. Micronutrient retention in the pouch-pack and especially in the UHT processed foods is generally higher than in canned products. Heat sterilization effects on nutritive values of proteins and carbohydrates are considered negligible, although the subject has not been studied extensively. Overall, heat sterilization is one of the most widely used food preservation techniques resulting in shelf-stable, nutritious foods that can be used to complement the seasonal fluctuations of the fresh food market.

10.5 OTHER FOOD PRESERVATION PROCESSES

In addition to drying, freezing and heat sterilization, other food manufacturing operations listed in Table 10–1 combine their manufacturing function with an element of preservation. The most common effect is that of thermal destruction of microbial populations. Combined with proper packaging or refrigerated storage, some of the relatively mild heat-processing treatments (pasteurization of milk, baking of bread,

FIGURE 10-9. The Tetra-brik packaging system for UHT sterilized milk. (Photo courtesy of Tetra-pak of Canada, Inc.)

smoking of processed meats) can result in a shelf-life of several weeks. Refrigerated storage in itself is a short-time preservation technique suitable for fresh as well as processed foods. However, some spoilage microorganisms can tolerate the typical refrigeration temperatures as shown by mold growth on yogurt, slime development on wieners, greening of processed meats, and other familiar signs of spoilage in foods kept in a home refrigerator for too long.

Preservation principles of other frequently used manufacturing techniques are based on altering the food environment and making it less suitable for microbial growth (Chapter 2). Preservation of jam, sweetened condensed milk, salted fish, jerky meat and similar foods relies on lowering of water activity, a_w, below approximately 0.75 by tying up the water with sugar or salt to make it unavailable for the microorganisms. Preservation of pickled vegetables, yogurt, sauerkraut and other fermented foods is accomplished by increased acidity to a pH which is below the optimum for most organisms, often below pH 4.0. Sometimes, the acid needed for the preservative action may be present naturally (as in fruits), may be added as vinegar or other food acidulant (as in salad dressings), or may be developed by microbial fermentation (as in yogurt).

The often controversial use of chemical additives (Chapter 9) is yet another processing technique aimed at retarding (not completely eliminating) microbial growth.

10.6 FOOD IRRADIATION

All approaches to food preservation used by the food industry today have been based on traditional proven concepts. A relatively recent technological advance in harnessing atomic energy for peaceful uses opened an entirely new possibility for food preservation by ionizing irradiation. Since the 1950s the process has been under thorough investigation so that enough research data are available on its safety, effectiveness, and technological advantages and limitations. The principle of irradiation is similar to that of canning: irradiation by γ-rays emitted by radioactive elements such as cobalt 60, or by a stream of electrons (β-rays produced by an electron accelerator) can be used to kill microorganisms present in food. The irradiation can be applied in low doses for selective inactivation of the more sensitive pathogenic or spoilage organisms only. The technical word *radurization* is being coined for this use to describe its relationship to pasteurization. High dose irradiation for complete sterilization of foods ("*radappertization*", indicating similar effects as in Appert's canning process) may be more limited for potential industrial use since different food materials respond differently to the high dose treatment. The main undesirable effects of sterilizing irradiation in some foods are related to various chemical reactions resulting in previously unknown flavor, color or texture defects. While the mechanisms of these changes are still poorly understood, the fact that a new processing technique may result in new and unexpected reactions is not surprising and only repeats the situation found with canning some 150 years ago.

Food irradiation is the only new food preservation principle that has emerged in the 20th century. The potential for its industrial utilization is significant. Because of its penetrating power, irradiation can be applied to whole pallet-loads of packaged foods (Figure 10-10); the process engineering advantages of this approach are self-evident. Irradiation is now used commonly for sterilization of medical disposable supplies. The application of irradiation in food processing has been thoroughly tested in many research institutes throughout the world. In the United States, the U.S. Army food laboratories at Natick, Mass. have been in the forefront of irradiation research for many years. In Canada, the Atomic Energy of Canada Ltd., one of the world's leading commerical suppliers of industrial irradiators, installed a production unit for potato sprout inhibition in the early 1960s. No health hazards or side effects resulting from the use of irradiated food have been reported for any of the foods tested, and *no radioactivity is induced in the food by this treatment.* Irradiated foods are now permitted in several countries on experimental or unrestricted basis (Table 10-8). In the U.S. and Canada, the industrial use of food irradiation was hampered in the past by the rather illogical regulatory classification of irradiation as a food additive. The Codex Alimentarius Commission (a

Table 10-8 International Use of Irradiation for Food Sterilization

FOOD ITEM	EFFECT OF IRRADIATION	PERMITTED IN
Onions, potatoes, carrots	Inhibition of sprouting	Holland, Belgium
Strawberries, mushrooms	Prolonged storage	South Africa, France, Holland, Hungary
Chicken, fish, ground meat	Control of *Salmonella*, prolonged storage	U.S.S.R., Hungary, Poland, Holland, W. Germany
Herbs and spices	Sterilization	Hungary, Belgium, Israel
Special hospital meals	Sterile food for special patients	England, Holland

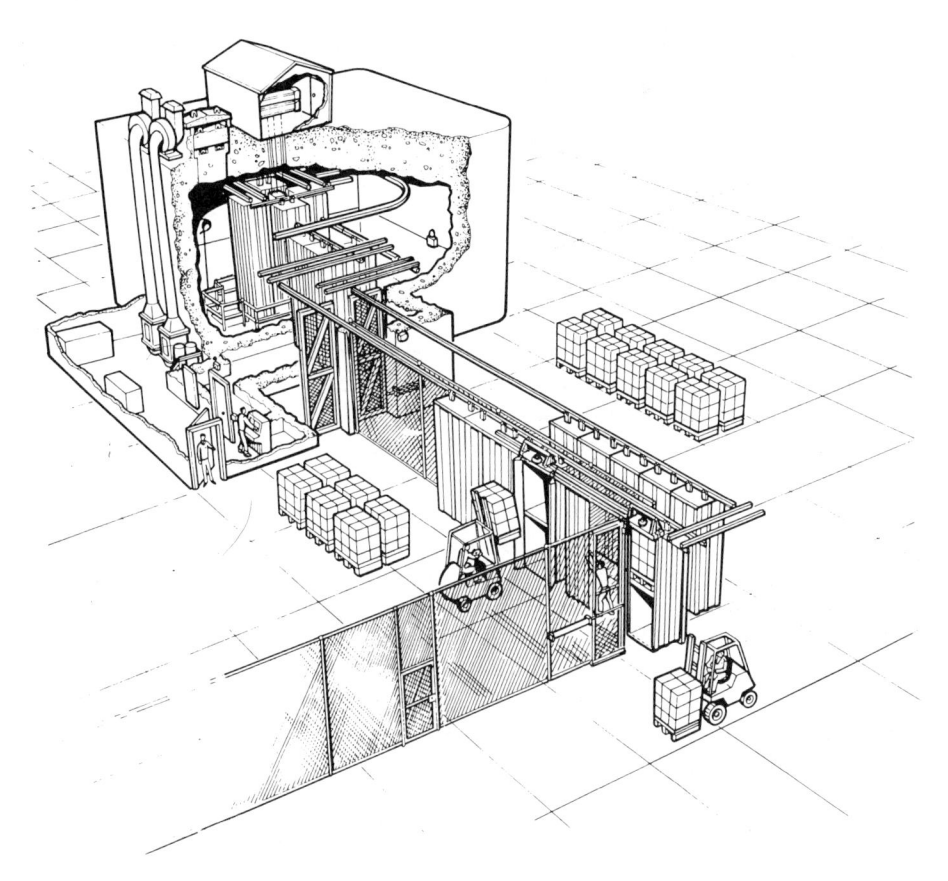

FIGURE 10-10. Food irradiation of palletized food products. (Courtesy of Atomic Energy of Canada Ltd., Industrial Irradiators Div.)

voluntary group of 122 countries affiliated with the United Nations and concerned with all international aspects of food processing, marketing and labeling) recently declared that "... it is self-evident that food irradiation is a physical process and, as such has most in common with other processes such as those based on heat, refrigeration, and drying.... Irradiation of any commodity (within a specified maximum irradiation dose) presents no technological hazard; hence toxicological testing of foods so treated is no longer required."

Since both Canada and the U.S. are prominent members of the Codex Alimentarius Commission, this agreement of international experts has paved the way towards the necessary regulatory change instituted in 1984 and potentially more widespread use of irradiation in North America. Since the new preservation technique offers significant energy savings, superior product quality for many types of foods, and important marketing advantages in prolonged shelf-life, it is safe to predict that its use in industrial food processing will significantly increase in the future.

10.8 NUTRITIONAL SIGNIFICANCE OF LONG-TERM FOOD PRESERVATION

The main objective of all food preservation techniques is to maximize the utilization of raw food materials. Since many raw agricultural commodities are produced on a seasonal basis, it is desirable to extend the availability of these popular consumer items well beyond their shelf life in fresh state. This will inevitably result in losses of certain micronutrients, especially vitamins that are known to be affected by long storage and by some of the processing technology. Thus, the primary goal of food processing for long-term preservation is the increased availability of macronutrients which are not greatly affected by any of the processing techniques used, and which are the predominant constituents of all foods. Some specifics of micronutrient losses of canned, dried, or frozen foods have been discussed in preceding paragraphs; generally, this is the price, albeit small, to be paid for year-round availability of many seasonal food items. While improvements in increased micronutrient retention of dried, frozen or sterilized foods may be possible, it should not be forgotten that hunger is caused by the lack of macronutrients, and that processing is a major factor in keeping these nutrients available.

10.9 CONTROL QUESTIONS

1. What are the effects of the three main food preservation techniques (drying, freezing, canning) on microorganisms? Will there be microbial spoilage in a) reconstituted fluid skim milk prepared from dry milk powder, b) previously frozen broccoli after thawing, c) a can of tomato soup after it has been opened, if these foods are left in a refrigerator after their handling? Explain each case.
2. Why is it that many dried foods exhibit flavors that are different from their fresh counterparts? Would you expect the same flavor problems occurring with frozen foods? Why or why not?
3. Why is roller ("drum") dried milk powder usually more yellowish than the same milk powder dried by a spray drier? Would you

expect any color difference between spray-dried and freeze-dried milk? Explain.

4. Calculate an approximate freezing point of an ice cream mix containing 17% sucrose, 7% lactose and, altogether, 39% total solids. Molecular weights of both sucrose and lactose are 342. Disregard freezing point depression effects of any other component of the mix.

5. Explain the principle of operation of a mechanical refrigeration system such as that used in a household refrigerator. In what respects is the mechanical refrigeration system different from the system using liquid nitrogen?

6. What will be the main difference between a process used for production of canned boysenberry juice (pH 3.1) and that for production of canned evaporated milk? Why?

7. Explain the technological principle of a direct UHT processing system and the package sterilization principle of the Tetra-pak system for packaging of UHT-processed foods.

8. What kind of food processing technique do the words "radurization" and "radappertization" describe? Are these words synonymous? What is the principle of the processing technique(s) used?

Suggested reference books—
Food preservation

Desrosier, N. 1977. *The Technology of Food Preservation.* Westport, Conn.: AVI Publ. Co.

Fennema, O. R., Powrie, W. D., and Marth, E. H. 1973. *Low Temperature Preservation of Food and Living Matter.* N.Y.: Marcel Dekker.

Goldblith, S. A., Rey, L., and Rothmayr, W. W. 1976. *Freeze Drying and Advanced Food Technology.* N.Y.: Acad. Press.

Hawthorn, J., and Rolfe, E. J. 1968. *Low Temperature Biology of Foodstuffs.* N.Y.: Pergamon Press.

Hoyem, T., and Kvale, O. 1977. *Physical, Chemical, and Biological Changes in Food Caused by Thermal Processing.* N.Y.: Appl. Sci.

Karel, M., Fennema, O. R., and Lund, D. 1975. *Physical Principles of Food Preservation.* N.Y.: Marcel Dekker.

Kessler, H. G. 1981. *Food Engineering and Dairy Technology.* Freising, Germany: Verlag A. Kessler.

Lopez, A. *A Complete Course in Canning*, 1982. Vol. I and Vol. II. Baltimore: The Canning Trade.

National Canners' Ass'n. 1968. *Laboratory Manual for Canners and Processors.* Westport, Conn.: AVI. Publ. Co.

Stumbo, C. R. 1973. *Thermobacteriology in Food Processing.* N.Y.: Acad. Press.

Toledo, R. T. 1980. *Fundamentals of Food Process Engineering.* Westport, Conn.: AVI Publ. Co.

Tressler, D. K., Van Arsdel, W. B., and Copley, M. J. 1968. *The Freezing Preservation of Foods,* Vol. I–IV. Westport, Conn.: AVI Publ. Co.

Van Arsdel, W. B., Copley, M. J., and Morgan, A. I. 1973. *Food Dehydration,* Vol. I and II. Westport, Conn.: AVI Publ. Co.

Van Arsdel, W. B., Copley, M. J., and Olson, R. L. 1969. *Quality and Stability of Frozen Foods.* N.Y.: J. Wiley.

11

Food Packaging Technology

11.1 PURPOSE OF FOOD PACKAGING

An integral part of all industrial food processing is packaging the processed food items. The subject of general packaging technology is so complex that it can be studied as a special technical discipline, separate from food technology. Design and operation of a given food processing or preservation technique, however, is often interrelated with proper choice of packaging. Manufacture of wieners, canned foods, sterilized milk and many other products would be impossible without packaging, which in these cases becomes an integral part of the *food processing technology* itself. In other cases, packaging serves mainly as a *convenience tool* for easy handling (liquid foods, cookies, nuts, etc.), portion packaging (fresh meats, cheese, butter), or even as a serving utensil (yogurt, T.V. dinners, marmalade jars). The traditional purpose of food packaging has been *protection* against deterioration due to damaging effects of our environment, especially contamination by microorganisms, dirt, unintentional chemical adulterants, as well as physical and chemical spoilage by light, oxygen, water, or drying out. More recently, packaging assumed a very important role of *information* (Chapter 9) and *content identification*, as well as that of a *marketing tool* including advertising, sales promotion, and various convenience features. The development of the modern supermarket food retailing system—as we know it in the Western world today—has been possible because of the *self-service* aspects of contemporary packaging. The virtual elimination of over-the-counter sales has had a very significant socioeconomic effect, resulting in profound lifestyle changes for a large segment of the society—the former shopkeepers—as well as for all of us, the consumers. Competition among food processors, one of the desirable features of a free-market economy, has been enhanced by supermarket packaging concepts.

Modern food packaging for retail is based on single serving or generally small-size packages. Most packaging materials used by the food industry today (Table 11-1) are still the same as those used in the past when many food items were packaged in bulk and portioned in neighborhood shops, often into containers supplied by the customers. This rather unsanitary concept of food merchandising, now abandoned in North America except for a few health (!) food stores, is still common in many less developed countries.

Food packaging materials may be differentiated according to the function they play and/or their interaction with the food itself. Those materials that come into direct contact with the edible portions of food, or the *primary* packaging materials, come under the special scrutiny of the food regulatory agencies (Chapter 3). By law, a primary food package must not yield to the food any substance that may be injurious to the health of the consumer.

Besides the primary packages used for containment of the food material itself, there are numerous occasions where *secondary* and even *tertiary* packaging is used. The primary food packages must often be assembled into a larger unit suitable for shipment, handling, or added protection. Examples of secondary or tertiary packaging include bags of potato chips in a box, plastic milk pouches in an overwrap or a crate, boxes of macaroni on a wooden pallet wrapped in a plastic shroud for easy loading and transportation, and many others.

Food materials in which the edible portion is naturally protected against the enviroment such as eggs, oranges, or bananas, are packaged with secondary packaging materials where regulatory control is much less stringent.

Table 11-1 Materials Used for Food Packaging

MATERIALS	EXAMPLES OF USE
Paper	Bags, boxes, cartons
Glass	Bottles, jars
Metal	Cans, aluminum foil
Plastics	Overwraps, pouches, cups, boxes, bottles
Laminates	Cartons for liquids, multilayered plastics
Wood	Crates, pallets
Cloth	Sacks, special packages
Wax	Coating of fruits and vegetables, some cheeses (Edam)
Edible Containers	Special uses (ice cream cones, cabbage leaves)

11.2 Properties and Uses of Food Packaging Materials   251

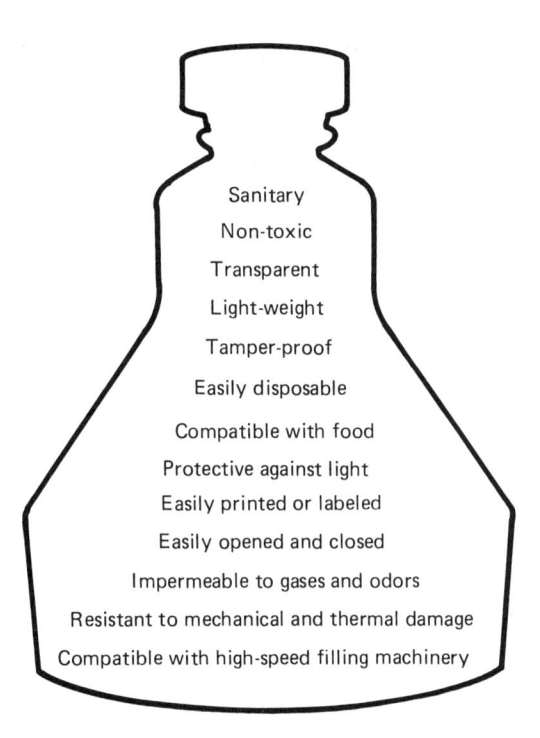

(Bottle-shaped diagram containing the following list:)

Sanitary
Non-toxic
Transparent
Light-weight
Tamper-proof
Easily disposable
Compatible with food
Protective against light
Easily printed or labeled
Easily opened and closed
Impermeable to gases and odors
Resistant to mechanical and thermal damage
Compatible with high-speed filling machinery

FIGURE 11-1. Properties of an ideal food container.

A food processor must understand at least the basic properties of the numerous food packaging materials available to provide the functions listed in Figure 11-1. It is apparent that some of the requirements are contradictory, e.g. light-permeable packaging material will not protect against light but will allow the consumer to see the content of the package. The choice of the proper material must be made by the food processor, based on food composition and its physical state (liquid, solid), knowledge of the various deteriorative reactions that might occur, intended storage conditions (including the time of storage), the socioeconomic situation of the anticipated customer, desired package attractiveness, the cost of the packaging material, the packaging technology selected, and the specific functional properties of packaging materials as listed in Table 11-2.

11.2 PROPERTIES AND USES OF FOOD PACKAGING MATERIALS

Four major materials—paper, glass, metal and plastics—cover practically all contemporary food packaging applications. Other materials listed in Table 11-1 such as wood, cloth or wax are nowadays of little

Table 11-2 Functional Requirements of Packaging Materials

FUNCTIONAL PROPERTY	SPECIFIC FACTORS
Gas permeability	O_2, CO_2, N_2, H_2O vapor
Protection against environmental factors	Light, odor, microorganisms, moisture
Mechanical properties	Weight, elasticity, heat-sealability, mechanical sealability, strength (tensile, tear, impact, bursting)
Reactivity with food	Grease, acid, water, color
Marketing-related properties	Attractiveness, printability, cost
Convenience	Disposability, repeated use, resealability, secondary use

importance for primary or secondary packaging needs. Each of the principal packaging materials is characterized by certain basic properties often determining its suitability for a particular food packaging operation.

Paper

In North America, paper is still the most popular food packaging material. The main desirable characteristics of paper are its low cost, versatility, ease of manufacture, excellent printability and light weight. Paper may be used both as a primary, and—perhaps more importantly—as a secondary container. Its protective properties against gases, odors, water, and grease are poor without various special treatments used by the paper manufacturing industry. There are various types of paper available for food packaging as described in Table 11.3.

To increase its protective properties while maintaining some of the advantages, paper is often combined with plastic films for good sealability and protection against liquid (milk cartons), or metal foil for increased gas, light and water vapor protection. These multilayered combinations (laminates) are further discussed in Section 11.3.

Glass

Probably the oldest packaging material for liquid foods, glass is still important in many situations today. Its main advantages are the total inertness against all physical or chemical effects of the food, impermeability to all gases and odors, good sealability combined with the often desirable ease of resealability, and suitability for a high-speed mechanical filling operation. Glass is manufactured from a molten

Table 11-3 Types of Paper Used As Food Packaging Material

PRODUCT	CHARACTERISTICS	EXAMPLE
Kraft paper	Brown, unbleached paper	Shopping bags
Bleached paper	White paper, may be glossy	White bags, printing paper, wrapping paper
Grease-proof paper	Very smooth surface	Some wrapping paper (meat products)
Parchment	Translucent paper treated with H_2SO_4 to gelatinize surface layers	Butter and margarine wrap
Glassine	Transparent, brittle	Overwraps on candy boxes
Paperboard, cardboard	Compacted paper pulp	Milk cartons, some fresh meat or vegetable trays
Corrugated cardboard	Two paperboard sheets interspersed with paper corrugations	Secondary boxes of many kinds

mixture of silica and potash with several other ingredients. Thus, its manufacture is cheap, in fact often cheaper than the rather expensive collecting, washing, and refilling operations involved in glass bottle recycling. The transparency of glass can be advantageous for aesthetic reasons or for content display purposes. Yet it can be modified by tinting, to offer a fairly good light protection. Labeling of glass with paper labels or direct printing is simple.

The susceptibility of glass to breakage (both due to mechanical impact and thermal shock), its heavy weight, and its nonbiodegradability are three major disadvantages that have diminished the importance of glass as the traditional package for some foods as milk, jams, or beer. For many other situations, though, the glass bottle is going to be the ideal package for a long time to come.

Metal

The single most important use of metal for food packaging is in canning. The total annual value of metal cans used by American food industry alone was estimated at over $4 billion—second only to the $7.8 billion spent for paper in all food packaging applications. The suitability of a metal can for heat processing is due to its good heat transfer properties, ease of manufacture and air-tight hermetic closure, complete protection against all gases, light and other environmental effects, and high resistance against pressure, mechanical, and thermal shocks. The steel used for the can manufacture, however, is heavy, expensive, and can react with the can content. For added protection, the steel used for can manufacture is coated with a thin layer of tin (even more expensive) or with other resistant compounds, enamels, and plastic coatings. Protective compounds are applied especially to both lids to prevent the combined effect of the food and the gas in the head space of a can.

Some of the canned foods (particularly fruits) can be very acidic and thus highly corrosive to unprotected metal. Faults in the protective layers may lead to a quality defect known as *hydrogen swell*—a bulging appearance of the can which is not caused by microbial spoilage but by the harmless chemical reaction between the food acid and the metal. Because of the container's rigidity, this defect, as well as swelling due to the various types of true microbial spoilage when gas is produced, could pose a significant danger for a person trying to open such a swollen can.

There are many types of food cans used by food industry today. Their identification, sometimes very confusing to the consumers, is based on the can diameter and height as shown in Figure 11–2, or on a traditional, industrial name identification. Some of the most typical cans are listed in Table 11–4.

Another increasingly important use of metal in food packaging is in

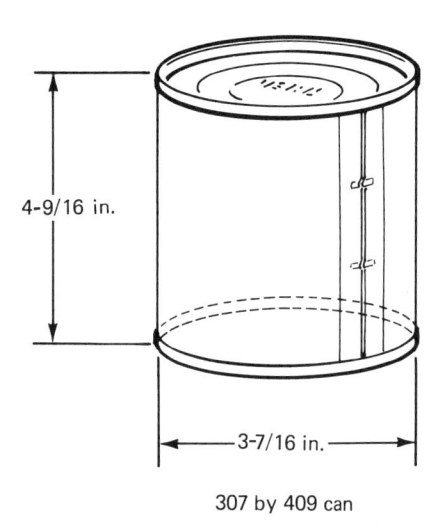

307 by 409 can

FIGURE 11–2. The numerical system for identification of sanitary cans.

Table 11-4 Trade Names, Dimensions and Typical Uses of Selected Cans for Retail

TRADE NAME	CAN SIZE[a]	CAPACITY (ml)	TYPICAL USE
No 1 – Picnic	211 × 400	310	Vegetables, soup, mushrooms
8Z Short	211 × 300	225	Fruit and vegetables
8Z Tall	211 × 304	247	Fruit and vegetables
8Z Mushroom	300 × 400	420	Mushrooms
No 300	300 × 407	433	Fruit juices, cranberries, spaghetti
No 1 Tall	301 × 411	474	Fruit, olives
No 303	303 × 406	480	Dry beans, fruit, soups, vegetables
No 2 Vac	307 × 306	418	Vacuum-packed vegetables
No 2	307 × 409	584	Fruit, juices, vegetables
No 2 Cyl.	307 × 512	750	Juices, soups
No 10	603 × 700	3,109	Various products

Note:
[a]Diameter × height (in inches). In both numbers, the first digit gives total inches, and the second two digits give sixteenths of an inch.

combination with paper and/or plastic in laminated films. These combination packages offer excellent protection afforded by the metal sheet (almost always aluminum foil) as well as good printability, heat sealability and other properties of paper or plastic. Metal containers from steel, aluminum or combination metals are also used for various pressurized aerosol cans, for large volume packaging for institutional food deliveries, and as beverage containers (beer, pop, and other beverage cans).

Plastics

Despite the proliferation of plastic food packages on supermarket shelves, these newest, most versatile and most adaptable packaging materials have barely started realizing their full potential. The base for numerous packaging films, tubs, bottles, pouches, trays and other food containers are a few synthetic organic chemical compounds, manufactured by *polymerization* of simple individual molecules into complex molecular chains. This gives the final material its molecular organization, orientation, stretchability, and the permeability properties important for various packaging tasks. In the chemical process of polymerization, the material properties desired for some special packaging functions can be controlled much more accurately than with any other food packaging material. Table 11-5 lists the most important plastics

Table 11–5 Examples of Basic Plastic Films Used in Food Packaging

MATERIAL (CHEMICAL NAME)	STRUCTURAL UNIT		IMPORTANT PROPERTIES
Cellulose	Cellophane	Glucose	Good strength, poor H_2O and gas barrier, good printability, no heat sealability
Polyethylene	–	Ethylene	Good heat sealability, excellent H_2O barrier, no printability
Polyvinyledene-vinyl-chloride	Saran	Vinyl	Excellent oxygen and moisture barrier, not very strong, heat sealable
Polyester	Mylar	Ethylene glycol + terephtalic acid	Excellent mechanical properties, poor heat sealability, fair H_2O and gas barrier
Polyamide	Nylon	Diamine + various acids	Excellent machinability, good strength, poor H_2O and gas barrier, heat-sealable

used as food packaging films, their structural base, and some of their properties of importance to the food processor.

A major technological advantage of virtually all plastic materials is their heat sealability and formability into desired structural configurations. These properties are related to the thermoplasticity of synthetic polymers. Various manufacturing techniques used by package manufacturers (specialized firms or the food processors themselves) are based mostly on melting or softening the granulated chemical feedstock ("resin") and solidifying it into the desired structures.

There are three principal processes available for package formation. In film *extrusion*, the molten plastic is transformed into a thin sheet upon forcing it through a narrow slit and stretching; in *injection molding*, the molten feed is injected into a metal mold where it solidifies into the desired shape; and in *blow molding*, compressed air is used to form the desired container inside a precision mold from a preformed mass of plastic feedstock. The blow-molding package formation may be combined with the subsequent food-filling operation in a single machine (Figure 11–3), as is increasingly common in packaging milk and other fluid foods. Similarly, processed sliced meat products or portion-packed pasty foods may be filled into tubs and trays formed by the vacuum-thermoforming technique from flat plastic films im-

FIGURE 11-3. Blow-molding filling machinery for packaging pasteurized milk. (Photo courtesy of Hoover Universal Co.)

mediately before filling (the "form-seal" principle). Yet another packaging system for liquid foods such as milk, employs the formation of flexible pouches from a flat plastic film by longitudinal and cross-sectional heat-sealed seams produced just before the milk is poured into the package (Figure 11-4). The newest achievement of the plastics-packaging industry is a plastic beverage can.

The widely varying permeability of plastic packaging materials to various gases may be one of the determining factors in selecting the most appropriate material. For some food packaging applications, films with high permeability to oxygen, but low permeability to water vapor, are desirable. In other cases, low permeability to nitrogen may be needed for foods packaged under nitrogen gas. Good resistance to acids, alkali, solvents, water and other food components is necessary for those plastics that are used as primary packages. Other properties that may influence the choice of a given packaging material by the food processor are printability (from virtually unprintable polyethylene to

FIGURE 11-4. Pouch-pack filling machinery for packaging milk.

excellent printability of cellophane), mechanical properties (strength, stretchability, resistance to puncture), and cost.

The economic unit of the plastic packaging films is their weight per roll. Thus, both the density and thickness of the film will play a major role in the final cost per unit of the food package. The conventional units of thickness used by the packaging trade are one *mil*, denoting a thickness of 0.001 inch, or 1 *gauge* (0.0001 inch). Typically, packaging films are several mils thick.

Tests for functional properties of all food packaging materials are available and constitute a major activity of food packaging engineers. In some food quality control procedures (especially in canning operations), frequent tests of packaging materials by the food processor are also required in order to avoid possible health hazards to consumers resulting from supplies of faulty packaging materials.

11.3 LAMINATED PACKAGING MATERIALS

As can be seen, there is no universal packaging material that would be ideally suited for all food packaging needs. To minimize the disadvantages of individual materials without sacrificing their desirable properties, a technique known as *lamination* is used to produce some of the most sophisticated packaging materials today. Although some of the earlier combination packages from paper could perhaps be termed laminates, the true era of laminated packaging materials has begun with the advent of plastics-packaging technology. Many flexible films used for food packaging are laminates in which two or more layers of various plastic sheets are combined to utilize the specific protective properties or other advantageous features offered by individual films. Two alternative processes—*extrusion* and *adhesion* lamination—may be used to seal together two or more layers of the packaging films, giving a combined property of all the components. As shown in Figure 11–5, the extrusion lamination is based on the thermoplasticity of the polymer resins (two films are fused together with a molten plastic material); in the adhesive lamination the two films are combined together with the help of an adhesive glue.

Typical laminates include combinations of easily printable, but water-permeable and heat-unsealable cellophane with other plastics; two or more flexible films with differential gas permeabilities, mechanical or other properties; and many multicomponent laminates of plastics with paper or metal foil. The plastic-paper combination is especially advantageous for fluid food packaging (dairy products, fruit juices, wine) since the sturdiness, printability and low cost of the paper carton is combined with the heat-sealability and water-resistance of

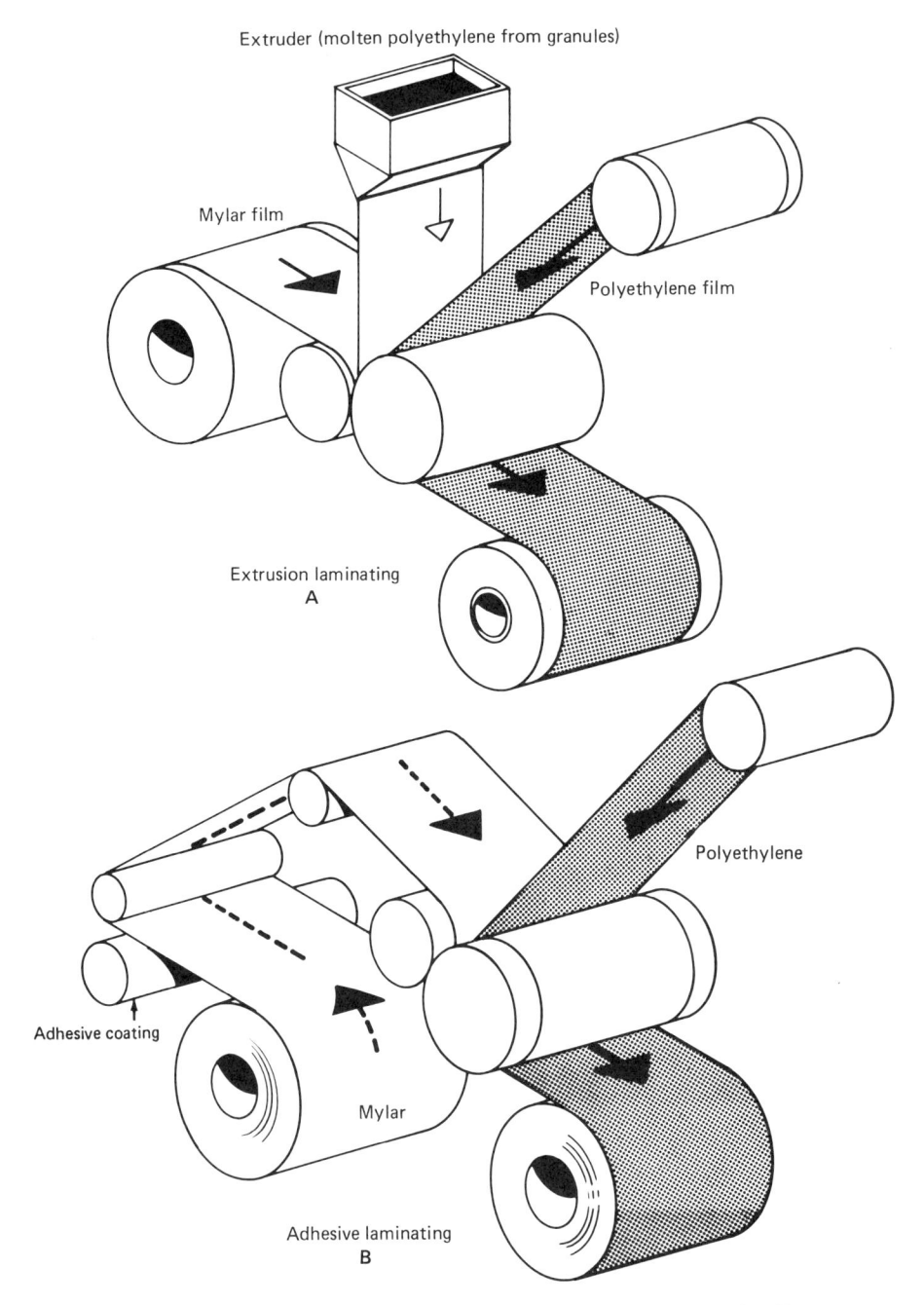

FIGURE 11-5. Schematic illustration of the extrusion (A) and adhesion (B) lamination processes.

the plastic. Some of the most complicated laminates are used for packaging of UHT sterilized food products (Chapter 10) which require adequate protection from the effects of light, minimum permeation of gases and total inertness of the package for many months. This is achieved by multilayered laminates combining paper, aluminum foil and several layers of plastic films (Figure 11–6).

Some of the recent significant developments in the food processing industry would not be possible without the availability of this modern packaging technology. The retortable pouch, the UHT processing of milk and other liquid foods, the "shrink-wrapping" technology used for container-shipments of many food items, even the moon explorations

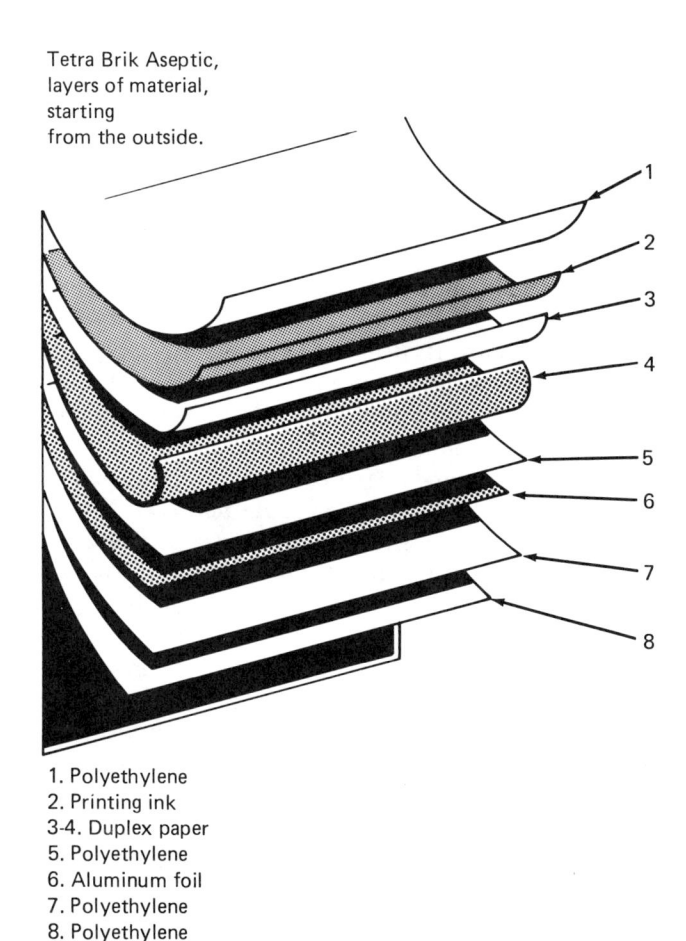

Tetra Brik Aseptic, layers of material, starting from the outside.

1. Polyethylene
2. Printing ink
3-4. Duplex paper
5. Polyethylene
6. Aluminum foil
7. Polyethylene
8. Polyethylene

FIGURE 11–6. Construction of the laminated packaging material used by Tetra-brik Aseptic Machinery. (Courtesy of Tetra-pak of Canada, Inc.)

by the U.S. astronauts, were all based on the developments in lamination technology.

11.4 EFFECTS OF PACKAGING ON FOOD STABILITY

Some of the requirements of package functionality are related to the various deteriorative reactions that may proceed in improperly packaged food. Thus, foods sensitive to oxidative fat deterioration (rancidity) may have to be packaged in a vacuum or an inert atmosphere. As in the packaging of whole dried milk under nitrogen, the packaging material chosen must be impermeable to oxygen as well as to inert gas. Dried foods must be kept in packages resistant not only to liquid water but also to water vapor, which could be absorbed from the surroundings especially by dry foods of hygroscopic nature (dried whey, tomato powder, fruit drink crystals). Water absorption could cause deterioration due to microbial growth if localized regions of high enough a_w are formed, or due to recrystallization of the moist sugars into a solid mass ("caking"). Appropriate packaging must be selected for foods containing light-sensitive components (e.g. riboflavin in milk, flavor compounds in beer, vitamin C in potato chips).

The requirement for protection against one agent may be combined with the desired permeability for another. Plastic materials used for fresh meat packaging should have high permeability for oxygen, permitting the oxygenation of myoglobin for the desired bright red color (Chapter 7), but low H_2O vapor permeability to avoid loss of weight by water evaporation. Differential permeability for CO_2 and O_2 may be required for packaging of fresh fruits or vegetables to accommodate their respiration requirements (Chapter 4). These complicated permeability factors can be satisfied mainly by using plastic packaging films that can be tailor-made for the permeability features desired.

All materials intended for protection against microbial invasion must be sealable and resistant against mechanical damage. Packaging systems used for sterile foods (metal cans, cartons for UHT processed foods, glass jars for baby foods, retortable pouch-packs) must be particularly effective in providing an airtight, hermetic closure to eliminate any microbial or other contamination from the surroundings after the process. The seals must be strong enough to withstand the rather severe heat treatment as well as the post-processing handling. Sterile foods are stored without refrigeration; any post-processing contamination due to improper packaging could result in spoilage or a severe health hazard if a single pathogenic contaminant grows without the usual competition of the mixed microflora of nonsterile foods.

Materials used for packaging of sterile foods or foods with prolonged non-refrigerated shelf-life must be sterilized, unless the complete food package itself is sterilized after the filling operation. Conventional sterilants used by UHT packaging machinery include hydrogen peroxide, heat, or ultraviolet light. A new promising technique for sterilization of food packaging materials is the ionizing radiation, ideally suitable for this purpose (Chapter 10). In the medical field, about 50 percent of all disposable instruments and other materials are now sterilized by γ-irradiation.

Many of the flexible packaging films produced by extrusion or blow-molding with hot air remain practically sterile until they are used in packaging machinery. The achievement of high sanitary standards by the modern food processing industry today has been facilitated by the gradual replacement of packaging systems with potential contamination risks, such as refillable bottles or primary paper packages, by the sterile or near-sterile flexible films.

11.5 MODIFIED ATMOSPHERE PACKAGING

Non-refrigerated shelf life of bakery products and certain other fresh foods can be extended by packaging these products in an atmosphere rich in CO_2 gas. In contrast to using N_2 (or CO_2) for packaging of dry products to avoid undesirable oxidative reactions, the CO_2 atmosphere in packages of fresh products retards substantially the microbial growth. Shelf-life of bread, jam-filled or cream-containing moist cakes and similar products can be extended several-fold thus giving the baker a significant advantage with respect to holiday and weekend sales. Other fresh foods where CO_2 packaging atmosphere has been used include cheese, cooked meats, bacon and other products with a potential for mold growth. The use of CO_2 is not suitable for fresh meats where the anaerobic atmosphere created by replacing O_2 by CO_2 would result in formation of the brown metmyoglobin pigment (Chapter 7), or in fruits and vegetables which require a careful CO_2 concentration control (Chapter 4).

The protective mechanism of the CO_2 atmosphere against microbial growth is not well understood. The theories proposed include creation of localized regions of low pH as the CO_2 dissolves in the water contained in the food, interference in the metabolic pathways of the microorganisms, and other possible effects. As the CO_2 inhibits growth of most microorganisms, aerobic as well as anaerobic, its effect is different from that of N_2 atmosphere or vacuum packaging which cannot be used to prolong shelf-life of fresh foods without refrigeration.

Good solubility of CO_2 in both water and fat may be advantageous for appearance of the packaged materials. As the gas is absorbed slowly, the package may show the tightness of a vacuum-packed product without the severe compression sometimes caused by vacuum packaging. The solubility of CO_2 in various food components must be kept in mind when using rigid containers as in packaging of soluble coffee, nuts, dried whole milk or biscuits. Partial vacuum caused by slow dissolution of the CO_2 atmosphere may cause distortion of the rigid containers. When flexible packaging films are used, their differential permeability to CO_2 and other atmospheric gases must be carefully evaluated. The use of CO_2 in food packaging is a relatively new packaging concept whose potential has not been fully explored yet.

11.6 ROLE OF PACKAGING IN THE FOOD DISTRIBUTION CHAIN

The modern food retail system would be inconceivable without the advances of food packaging. Many of our most comon foods could not be manufactured if proper packaging technology would not be available. Our comfortable lifestyles would be threatened if we had to spend a significant portion of our time (and/or money) shopping daily for perishable foods. Many foods would be highly seasonal. Contamination by insects, microorganisms, or environmental pollutants could result in nutrient losses and health hazards as commonly experienced in the past. In short, the remarkable achievements of the modern food processing industry would not be possible without equally remarkable advances of the food packaging technology.

It can be argued that a whole new system of food delivery from the producer to the consumer has evolved with the concept of portion-packing. Virtually all our foods are now packaged, occasionally to the point when the question of overpackaging arises. The relative costs of food packages are sometimes substantial when compared with the value of the food contained in them. In some instances, such as the single-serving packs of mustard, ketchup or salt, the reason is our new lifestyle; in other cases, it's convenience or marketing considerations.

From the standpoint of consumer safety, however, we are much better off with overpackaged rather than underpackaged foods. In absolute terms, the overall food packaging costs are only a fraction of the already low food costs. Compared to the benefits that the food distribution chain offers to all of us, the costs of packaged foods are a bargain.

Technological achievements of the modern food processing and packaging industries have created the most sophisticated food delivery system in the history of mankind. In North America we can buy the most varied, nutritious, safe and enjoyable food at lower cost than anywhere else in the world. We have eliminated nutritional disease, malnutrition, and famine. We live longer to enjoy life and one of its main pleasures—our daily food.

The food industry has come a long way.

11.7 CONTROL QUESTIONS

1. What is the most important function of the packaging technology used for the following products: a) UHT-processed sterilized chocolate milk in a laminated carton; b) ice-cream stick novelty in a paper overwrap; c) potato chips in an aluminum foil-laminated bag; d) wine in a "bag-in-box" type, four liter carton "cask"; e) tea biscuits in a decorated metal box?
2. Are there any particular legal requirements that must be satisfied before a new plastic material can be used for food packaging? Is it the same for paper? Metal?
3. In a product development project, consideration is being given to two alternative packaging materials. Both materials have excellent water vapor barrier properties and fair oxygen barrier properties. To achieve equivalent O_2 protection, material A must be used at a 5 mil thickness, while material B gives the same protection at 2 mil thickness. Material B costs three times as much per mil of thickness as material A. The tensile strength of material A is 1.5 times that of material B at an equivalent thickness. If there are no other constraints, which material would you use and why? What other important material properties may influence the final selection?
4. What is hydrogen swell and how can it be prevented?
5. The multilayered laminated packaging material used for packaging UHT-processed products consists of three main components—paper, aluminum foil and plastic. If you were to design a new packaging material consisting of only one layer of each of these three materials, what would be their arrangement starting from the outside? Explain the function of each of these three materials in the package.
6. If you want to prolong the shelf-life of carrots from your own backyard garden, you can store them at refrigerated conditions in a plastic bag. Should the plastic have high permeability to H_2O? CO_2? O_2? Why or why not?
7. Suggest a suitable packaging technology for the following protective requirements: a) protection of light-sensitive vitamin B_2

(riboflavin) in a whey drink, b) protection of dried onion powder against caking, c) protection of dried coffee powder against oxidation, d) protection of a semi-moist bakery product against microbial spoilage. In each case, explain briefly the problem and why would your selected technology be suitable for prevention of the problem.

Suggested reference books—
Food Packaging

Brody, A. L. 1970. *Flexible Packaging of Foods.* Cleveland: CRC Press.

Crosby, N. T. 1981. *Food Packaging Materials. Aspects of Analysis and Migration of Contaminants.* Barking, Essex, England: Applied Sci. Publ.

Hanlon, J. F. 1983. *Handbook of Package Engineering.* New York: McGraw-Hill Book Company.

Modern Packaging Encyclopedia. New York: McGraw-Hill Book Co.

Sacharow, S. and Griffin, R. C., Jr. 1973. *Basic Guide to Plastics in Packaging.* Boston: Cahners' Books.

Sacharow, S. and Griffin, R. C., Jr. 1980. *Principles of Food Packaging.* Westport, Conn.: AVI Publ. Co.

The Packaging Encyclopedia. 1984. Des Plaines, Ill.: Cahners' Publishing.

Appendix I
Procedures for Mass and Energy Balance Calculations

A.1.1 MASS BALANCE

There are several simple procedural steps that should be followed in a mass balance calculation:

1. Define your operation for which the calculation is made (i.e. sausage mixer; milk evaporator in combination with drier; sherry blender; etc.). It is very helpful to draw a schematic picture. Identify the desired results (weight of components to be mixed, amount of water to be evaporated, etc.).
2. Select your *basis* for calculation (i.e. 100 kg of final sausage mix; 1 L of sherry; 100 lb. of milk to be dried, etc.).
3. Write one linear equation for every *material component of interest* in the system, balancing weights of all materials or their components entering the system against all materials or components leaving the system. One equation should always be the *total* mass balance; i.e. "all materials in = all materials out."
4. Solve the set of linear equations for all unknowns (typically, these are the desired weights of individual ingredients) *and check* the solution by substituting the results into the initial equations (of course you should end up with equalities).
5. Write an *appropriate* answer!

   The materials in the individual equations can be either food products (i.e. wine A, wine B, wine C, etc. for sherry) or components of these materials, i.e. protein, water, alcohol, etc. The following examples should illustrate the whole procedure.

Problem 1

Frozen carrots-and-peas mixtures from two different suppliers are being used in the manufacture of canned beef stew. The mix from Supplier A contains 80% carrots and 20% peas; mix B contains 30% carrots and 70% peas. How much of each of the two mixes must be blended together to be used for a product which should have equal amounts of peas and carrots?

Step 1: Define the operation.

```
Mix A →  ┌──────────┐
         │          │
         │ Blender  │  → final product
Mix B →  │          │
         └──────────┘
```

Our operation is *blending*, and our desired results are the amount in kg of mix A and mix B to be blended together.

Step 2: Select a basis for calculation. Let's decide to base our calculations on 1,000 kg of the final peas-and-carrots mix. (We could just as well use 1 g of mix A, or 5 tons of mix B; the corresponding equations would be numerically different but the final *relative* answer should be the same in all cases.)

Step 3: Write two linear equations containing the two unknowns, the amounts of A and B to be mixed. Conveniently, one of these two equations will be the total material balance. Use the symbols of mixes to denote their unknown quantities.

Equation 1. *Total* material balance, balancing *total* mass of everything, in kg against kg, without differentiating the components:

A kg + B kg = 1,000 kg   (total mass - Equation A–1)

(NOT = 1,000 (A+B) or = 100 C+P or anything like that on the right side of the equation!).

Equation 2. We need another equation since we have two unknowns. We shall decide to write a material balance for carrots. Since there is 80% carrots in the A kg of the mix A, 30% carrots in the B kg of the mix B, and there should be 50% carrots in the 1,000 kg of the final mix selected as a basis for our calculations, we shall write the equation for carrots as

A.1.1 Mass Balance

$$0.8 \times A \text{ kg} + 0.3 \times B \text{ kg} = 0.5 \times 1{,}000 \text{ kg} \qquad \text{(Equation A-2)}$$

(Note that we do not use the symbols C and P; all we need to know is the carrot content in each unit which is given by the percent fraction multiplied by the weight of the specifix mix. It is helpful to identify each equation as to the kind of mass for which it is written, i.e. total, peas, carrots, etc.) Since we have only two ingredients in our system (mix A and mix B), we do not need another equation for the peas. (We could have used the equation for peas instead of that for carrots or the total mass equation.) If we were to have more components in the final mix (i.e. beans, corn, etc.) and more mixes to use, we would need more equations.

Step 4: Solving these two equations for A and B will yield the amount of each of the two *mixes* (A and B) to be used in the final blend. There are various algebraic procedures available to accomplish the task; these can be found in any linear algebra textbook. The general method, applicable to all linear equation systems, is based on expressing one unknown in terms of the others and thus reducing the system of equations to one equation only. Thus, from Equation A-1:

$$A = 1{,}000 - B \qquad \text{(Equation A-3)}$$

This can be substituted into Equation A-2:

$$0.8 \times (1{,}000 - B) + 0.3 \times B = 0.5 \times 1{,}000 \qquad \text{(Equation A-4)}$$

Equation A-4, which contains only one unknown, is now solved for B:

$$\begin{aligned} 800 - 0.8B + 0.3B &= 500 \\ 0.5B &= 300 \\ B &= 600 \end{aligned} \qquad \text{(Equation A-5)}$$

The value for B is substituted back to Equation A-1 (this is simpler than Equation A-2 which could be used as well).

$$\begin{aligned} A + 600 &= 1{,}000 \\ A &= 400 \end{aligned} \qquad \text{(Equation A-6)}$$

As shown in Equations A-5 and A-6, the results in this particular problem are A=400 kg, B=600 kg. The check using Equation A-2: 400 kg × 0.8 = 320 kg carrots from mix A; 600 × 0.3 = 180 kg carrots from mix B, together there are indeed 500 kg carrots in the final mix.

Step 5: The appropriate answer: We have to blend 400 kg of mix A and 600 kg of mix B *to produce 1,000 kg of the desired final mix*. (This is important to state since the actual batch weight may be different and the results must be adjusted. For example, if we want to actually produce only 500 kg of the final product, we will mix 300 kg of mix B and 200 kg of mix A.)

Problem 2

An orange juice concentrate is to be made by evaporating water from fresh orange juice. The fresh juice contains 12% total solids (T.S.), and the concentrate should contain 42% T.S. How much water must be evaporated in the process?

Step 1. Define the operation.

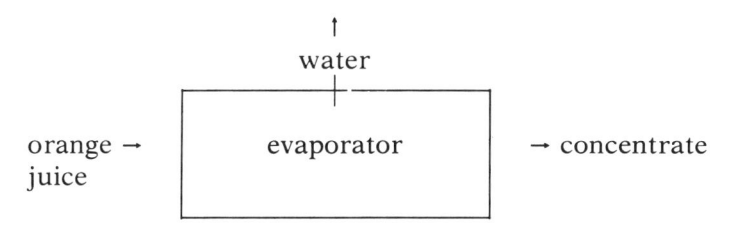

```
                          ↑
                        water
                          |
    orange →    ┌─────────────────┐
      juice     │    evaporator   │  → concentrate
                └─────────────────┘
```

Step 2: Select a basis (remember, this is our decision). We shall base this calculation on 1 kg of incoming orange juice.

Step 3: Set up the linear equations:
Equation 1. Total material balance (kg mass against kg mass)

1kg (juice) = W kg (water evaporated) + C kg (concentrate), mathematically, 1 (kg) = W (kg) + C (kg)

Equation 2. Balance of *total solids* (kg T.S. against kg T.S.)

1×0.12 (kg T.S.) = $W \times 0$ (kg T.S.) + $C \times 0.42$ (kg T.S.)

Step 4: Solve the system of linear equations. (In this case, the solution is trivial, since Equation 2 has in fact only one unknown; the results are

C = 0.285 kg and W = 0.715 kg

Step 5: Appropriate answer:- From every kg of fresh orange juice we can manufacture 0.285 kg of the concentrate, and thus must evaporate 0.715 kg of water.

Things will become more complicated when more than two components are to be accounted for. However, the principle is exactly the same, i.e. we need as many equations as unknowns that we are dealing with.

Problem 3

A speciality sausage mix is produced by blending meats A, B and C (e.g. mutton trimmings, beef head, pork jowl). The composition of these materials (in those components that are of interest to us) and the composition of the desired sausage are as follows:

Ingredient	% Protein	% Fat
A	12.5	35.5
B	3.8	50.8
C	17.0	15.0
Desired Mix	13.0	30.0

Step 1: Define the operation.

$$A \rightarrow \boxed{\text{mixer}} \rightarrow \text{final mix}$$
$$B \rightarrow$$
$$C \rightarrow$$

Step 2: Select a basis.
Let's say that our blender holds 500 kg maximum - i.e. we are producing batches of 500 kg. This will be our basis for calculation.

Step 3: Set up the equations.

Equation 1. A (kg) + B (kg) + C (kg) = 500 (kg)

Equation 2. (for protein):
$\quad$ A × 0.125 (kg) + B × 0.038 (kg) + C × 0.17 (kg) = 500 × 0.13 (kg)

Equation 3. (for fat):
$\quad$ A × 0.355 (kg) + B × 0.508 (kg) + C × 0.15 (kg) = 500 × 0.3 (kg)

Step 4: Solve the system of equations. The general method can again be used; two substitutions will be necessary to first reduce the system of three equations in three unknowns to two equations with two unknowns and finally to one equation with one unknown. A more rapid technique, such as the use of matrices, computerized programs, etc., are used in practice.

Step 5: Answer: To prepare a batch of 500 kg of the desired mix, we must blend 249.6 kg of ingredient A, 66.5 kg of ingredient B, and 183.9 kg of ingredient C.

In the relatively rare instances where only *two* components are being *mixed together,* a quick and simple procedure called Pearson's

272 Procedures for Mass and Energy Balance Calculations

square can be utilized instead of the above calculations. The procedure is especially suitable if one needs to determine *ratios* at which the two components must be mixed together. The procedure can be best illustrated by the way of the following example.

Problem 4

Regular homogenized milk (containing 3.5% butterfat) and light coffee cream (containing 20% butterfat) are to be mixed to produce "half-and-half" cream of 10% butterfat. In what proportions should the two components be mixed?

Step 1: Draw a rectangle (an analogy for defining the operation in the above procedure). Label the *horizontal* lines with the names of the two products to be mixed:

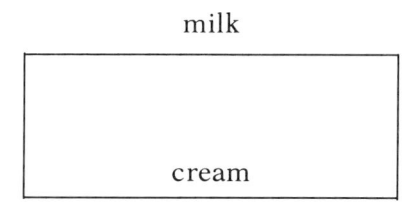

```
        milk
   ┌─────────────┐
   │             │
   │             │
   │    cream    │
   └─────────────┘
```

Step 2: Enter the appropriate compositional information *of interest* (i.e. the percentage butterfat in this case) in the left hand corners of the two component lines, and the desired composition of the final product in the center of the box:

```
           milk
   ┌─────────────────┐
   │ 3.5             │
   │         10      │
   │ 20              │
   │        cream    │
   └─────────────────┘
```

Step 3: "Mix the two components" by crossing diagonally through the center figure; subtract, in each case, the larger figure from the smaller one along the diagonal line. Enter the result in the right hand corners.

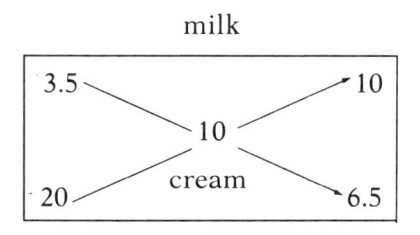

```
           milk
   ┌─────────────────┐
   │ 3.5╲       ╱10  │
   │      ╲ 10 ╱     │
   │      ╱    ╲     │
   │ 20╱  cream ╲6.5 │
   └─────────────────┘
```

Step 4: Read the results *in parts* (by weight) of the appropriate two components to be mixed—i.e. we need 10 parts of homogenized milk (top line) and 6.5 parts of the coffee cream (bottom line). The result is also valid if we have a continuous mixing device; the ratio of the two streams being mixed must be 10:6.5. A simple conversion of the ratio to any desired total batch weight would give the actual amounts (in kg) of the two materials to be used.

When more than two materials are being mixed, or when the final product composition is defined in more than just one component (i.e. protein and fat could both be required), the Pearson's square cannot be used. The regular material balance approach, often computerized, is the most widely used procedure in the food industry for these types of calculations.

A.1.2. ENERGY BALANCE

The procedure for calculating energy balance is based on a similar approach used for mass balances; all energy entering the system must be accounted for in the various forms in which it leaves the system. Usually only one equation is needed, since there are no unknown energy "ingredients" that are being mixed together. The main complication in energy balances is the need for various material properties data (specific heats, latent heats of vaporization and fusion, freezing or boiling points, as well as process-related data, especially temperatures). A simple example will provide an illustration of the procedure; the details are beyond the scope of this book.

Problem 5

A freezing line for sweet corn is being designed for a desired capacity of 7,000 kg/hr. How much heat will have to be removed during the freezing if the corn enters at 20°C and leaves the freezer at −10°C?

Step 1 - Define the operation: The operation is freezing, and the desired result is the total amount of heat energy to be removed.

Step 2 - Basis for calculation: In this case, the basis is provided in the problem as 7,000 kg/hr of corn.

Step 3 - Write the energy balance equation: The equation will balance the total heat (Q) leaving the system against all the heat forms entering the system; in this case, the heat load is represented by the unfrozen corn entering the freezer. However, during the freezing process the corn is first cooled to its freezing temperature (Q_1), frozen at this temperature by removing the latent heat of solidification (Q_2), and further cooled in the frozen state (Q_3). Thus, the energy balance equation is

$$Q = Q_1 + Q_2 + Q_3 \text{ (kJ/hr)} \qquad \text{(Equation A1-7)}$$

Step 4 - Solve the energy balance equation: To calculate the Q_1, Q_2, and Q_3, we must know the specific heat of unfrozen corn (C_1), the latent heat of solidification (R), the specific heat of frozen corn (C_2), the freezing point of corn T_F, and its moisture content M. These data can be found in literature; thus

$$C_1 = 3.35 \text{ kJ/kg°C}$$
$$C_2 = 1.80 \text{ kJ/kg°C}$$
$$R = 251 \text{ kJ/kg}$$
$$T_F = -1°C$$
$$M = 75\% \text{ (w/w)}$$

From these data, we can calculate how much heat each of the three steps in the freezing process represents. At first cooling 7,000 kg of corn from 20 to $-1°C$ will evolve

$$Q_1 = 7,000 \text{ kg/hr} \times 3.35 \text{ kJ/kg °C} \times 21°C = 492,450 \text{ kJ/hr}$$
$$\text{(Equation A1-8)}$$

Similarly, solidification of all the water contained in the 7,000 kg of corn at $-1°C$ without any change in temperature will give

$$Q_2 = 7,000 \times 0.75 \text{ kg/hr} \times 251 \text{ kJ/kg} = 1,317,750 \text{ kJ/hr}$$
$$\text{(Equation A1-9)}$$

And finally, cooling the frozen corn to $-10°C$ will require removing additional heat:

$$Q_3 = 7,000 \text{ kg/hr} \times 1.8 \text{ kJ/kg °C} \times 9°C = 113,400 \text{ kJ/hr.}$$
$$\text{(Equation A1-10)}$$

Balancing the total of the equations A1-8, A1-9 and A1-10 against the unknown—the amount of heat to be removed (or the amount of refrigeration that must be available) gives

$$Q = Q_1 + Q_2 + Q_3 \text{ (kJ/hr)}$$
$$Q = 492,450 + 1,317,750 + 113,400 \text{ (kJ/hr)}$$
$$Q = 1,923,600 \text{ kJ/hr} \qquad \text{(Equation A1-11)}$$

Step 5 - Appropriate answer: The freezer must be designed with refrigeration capacity capable of removing at least 1,923,600, or about 2,000,000 kJ/hr. The actual capacity must be higher as we did not consider any additional heat input from the surroundings into the freezing process. The magnitude of this "refrigeration loss" will depend on proper insulation of the freezing tunnel; however, in a more detailed design engineering calculation the loss must be taken into account.

Suggested reference books—
Mass and Energy Balances:

Charm, S. E. 1978. *The Fundamentals of Food Engineering.* Westport, Conn.: AVI Publ. Co.

Earle, R. L. 1983. *Unit Operations in Food Processing.* Oxford, England: Pergamon Press.

Toledo, R. T. 1980. *Fundamentals of Food Process Engineering.* Westport, Conn.: AVI Publ. Co.

Appendix II
Selected Literature Topics

The purpose of this section is to provide supplementary reading related to several subjects which were discussed in the text. These articles, reprinted here with the permission of the original publishers, address some aspects of the contentious relationship between human health and nutrition vs. modern agricultural and food technology. The articles are provided in their entirety and without any commentary thus representing the views of the authors. It is hoped that this section will be useful for initiating discussion in the classroom setting, as well as for other readers ready to agree—or argue—with the views of others.

SOLE FOODS AND SOME NOT SO SCIENTIFIC EXPERIMENTS

D. M. Hegsted, Ph.D., and Lynne M. Ausman, Sc. D.

Elaborate experiments have been devised for the purpose of showing that one or another popular food is, in fact, toxic. The authors have duplicated these "scientific" experiments and drawn several interesting conclusions useful in responding to some of the popular nutrition science in the lay press.

The abundant and relatively inexpensive food supply which we enjoy in the United States is due to the success of modern agriculture. The production, harvesting, processing, and distribution of this food requires a system that is quite different from that which was possible when people relied largely upon food produced in their own locality. An increasingly large proportion of the food we eat must be processed in order to preserve it and make it available to the consumer. Most American consumers demand and are willing to pay for as much freedom as possible from the work of preparing food in their own kitchens. The wide acceptance of prepackaged, convenience foods is adequate evidence that few housewives are willing to pluck the feathers off chickens, bake their own bread, or spend as much time in the kitchen as their mothers did.

It is clear that if a large proportion of our food supply is preprocessed and prepackaged, the housewife has little control over the nutritional content of the diet her family eats. It is also clear that the availability of many of these foods in the market depends upon the use of various materials such as food additives in their manufacture. Since food processing is a rather recent development, it is not surprising that we have not had time in which to learn the actual long-term consequences of living on a diet of such foods. Vigilance is, of course, required. However, there is no reason to assume *a priori* that living off processed food is detrimental to health. Indeed, if we know what should and what should not be in foods and how much of each is permissible, a food supply of this nature provides the opportunity to "tailor make" and actually improve the nutritional quality of our diets.

Concern over the complex problems involved in providing an adequate and safe food supply for all Americans is legitimate, but it has led to many ill-conceived and illogical recommendations, many of which could be dismissed if they had not misled so many people. Apparently experiments labeled "scientific" are the most impressive to the lay reader who is unable to distinguish between the good ones and bad ones. We wish to present here the results of a "scientific experiment" similar to some of those that have been presented to the public in recent years and have been widely discussed. It is hoped that the reader will be able to discern the illogical conclusions that have been drawn from such experiments.

A popular type of research consists in feeding one food to test animals until they die, then crying, "Ah ha! So and so will kill you." The conclusions may be right, but in too many cases the experiment is wrong. The results, taken out of context or improperly interpreted, could be entirely misleading. We tested a number of single foods commonly consumed by the American public, none of which proved capable of supporting normal growth and development in young rats!

Young growing rats are the most popular species for nutritional experiments. This deserves some comment. The nutritional needs of rats are better known than those of any other species. As a model for man, nevertheless, it should be noted that the species has serious deficiencies. It is instructive to note that a well-nourished child of 2 to 3 years of age who consumes about 300 gm of food per day, weighs about 14 kg (31 pounds) and gains from 4 to 6 gm per day. Five gm of body tissue contain about 1 gm of protein, and the child's diet must provide enough protein to maintain his body structure and deposit an additional gm of new protein each day. A young rat weighing 50 to 100 gm (0.1 to 0.2 pounds) and consuming only 10 to 15 gm of food per day, also gains about 5 gm a day. Thus, this little animal must make the same amount of new tissue protein every day as the child, but its food intake is much smaller. The level and quantity of protein in the diet is enormously more important for the young rat than for the young child. Many foods that are seriously inadequate in protein for the young rat are less so, or perhaps adequate, for the young child. In this regard, we should note that there is little dispute over the fact that, although only 6 to 7 percent of its calories are protein, breast milk is an adequate and appropriate food for babies. This level of protein is not adequate for young rats. Because of its rapid growth, the young rat also has greater needs for several other nutrients than do infants and children.

On the other hand, a young rat will grow very well on a diet that contains no vitamin C, folic acid, or some other nutrients which are essential for man. The fact that a young rat fed on a particular diet shows good growth is no assurance that the diet will also be adequate for the human species. It is clear that the results of an experiment must be interpreted in the light of what is known about the nutritional requirements of the species used and of the nutritional requirements of man.

Finally, it should be noted that growth experiments in rats provide no information about the long-term effects of various diets. Williams, Hefley and Bode (*Proc. Natl. Acad. Sci.* Vol. 68, p. 2361, 1971), after feeding various foods to young rats, concluded that "eggs proved to be a remarkably complete food." These authors ignore the abundant evidence that, for man, coronary heart disease is a major health problem that is associated with elevated plasma cholesterol levels and that plasma cholesterol levels are partially determined by the dietary level of cholesterol. Furthermore, the rat is little affected by dietary cholesterol and is very resistant to atherosclerosis. Almost everyone agrees that it is wise for Americans to limit their consumption of saturated fat and cholesterol. Thus, conclusions based upon the rat can be misleading.

Sole Foods

One 13-week experiment involved 54 young male rats which were obtained from a commercial source. They had previously received a good diet and weighed, on the average, 75 gm. They were divided into groups of 6 animals each of approximately the same average weight. Each group then received water and one of the following commonly-acknowledged nutritious foods as its sole diet:

Group 1–a commercial dog food which long experience has shown to allow good growth in young rats and dogs. We consider this the normal or "control" group.
Group 2–whole pasteurized milk.
Group 3–hamburger steak, medium-well-done.
Group 4–commercial skim milk powder.
Group 5–commercial enriched white bread.
Group 6–one of the highly enriched breakfast cereals.
Group 7–frozen french fried potatoes, cooked according to the instructions.
Group 8–frozen orange juice concentrate.
Group 9–fresh spinach.

All of the animals were weighed twice weekly and their condition noted. Animals that became terminally ill were anesthetized with ether and killed.

The reader might test his nutritional knowledge by guessing what happened to the animals in the various groups.

As expected, those receiving the commercial dog food grew very well and remained normal in all respects. This was not true of any of the other groups.

The most striking result was that none of the animals fed fresh spinach survived more than 3 days. The cause of these casualties was probably the rather high oxalic acid content of spinach, since this precipitates calcium as an insoluble salt. The kidneys were not examined microscopically, but it is known that diets containing excessive oxalate cause crystals of calcium oxalate to form in the kidneys, blocking their usual function.

The growth of the other groups of animals varied. Those on the milk diet showed a rather satisfactory rate of gain during the first several weeks but ceased to gain after approximately 9 weeks, while those consuming only orange juice showed a loss of weight during the entire experiment. Of the group on the orange juice and that on the hamburger diet, only one animal survived the 13 weeks of the experiment. Two of the 6 animals fed skim milk powder died during this time.

In addition to the observations on weight, it should be noted that the rats fed only whole milk became pale and anemic; those fed hamburger became severely paralyzed; those fed nothing but skim milk soon became blind with cataracts. All the animals maintained on white bread, enriched breakfast cereal, french fried potatoes, or orange juice showed loss of hair to varying degrees and had an unkempt appearance.

Milk is known to be a relatively complete food, except that it is a very poor source of iron. In fact, feeding a whole-milk diet is one of the classic ways of producing iron deficiency for experimental purposes. Therefore, it was not unexpected that the animals living on milk became iron-deficient and developed anemia.

Skim milk powder is similar to whole milk, except that the fat has been removed. This process, it should not be forgotten, also removes the fat-soluble vitamins. Thus, skim milk is much lower in calories than whole milk and lacks the vitamins A and D in whole milk. However, the development of cataracts in the rats fed skim milk resulted from the high content of lactose in this diet. Lactose is the sole source of carbohydrate in milk. It breaks down into two simpler sugars, glucose and galactose, in the body. Since skim milk contains no fat, considerably more of it than of whole milk must be consumed to obtain the proper calorie supply. This high consumption overloads the capacity of the body to metabolize galactose, which accumulates in the body tissues and causes cataracts. This phenomenon has been known for many years. However, there was a report in the June 12, 1970 issue of *Science* that the feeding of yogurt to rats also caused cataracts and blindness. Characteristically, this was duly reported in *The New York Times* and elsewhere with the implication that there is something wrong with yogurt and it might be a dangerous food.

Although the animals forced to live on hamburger alone grew well initially, every one of them eventually developed paralysis. This also was expected. Meat is very low in calcium so, as these animals grew, they had to fall back on the calcium already present in their bones. The depletion resulted in longer, but very thin and fragile, bones that eventually collapsed and caused the paralysis. It is of some interest to note that meat is very high in the essential nutrient, phosphorus, but that this is a disadvantage rather than an advantage, because the high intake of phosphorus inhibits the utilization of the small amount of calcium that meat does contain.

The growth performance of the remaining groups, those receiving white bread, breakfast cereal, french fried potatoes, and orange juice was most likely influenced by the quantity and quality of protein each food contained, although their content of other nutrients varied greatly. For example, the animals on white bread grew better than those on breakfast cereal, even though the breakfast cereal had relatively high levels of many of the vitamins added to it. Those fed orange juice alone could not possibly grow, because their diet contained practically no protein. Orange juice is a good source of vitamin C but since rats, unlike man, monkeys, and some other species, can synthesize vitamin C in their bodies, this nutrient is of no use to them. Of all the foods fed, only spinach and orange juice contain significant quantities of vitamin C. It is pertinent to point out, that had this experiment been done with a species requiring vitamin C in the diet, the outcome of these experiments would have been quite different.

It should also be noted that although hamburger, white bread, french fried potatoes, and skim milk are all poor sources of vitamin A, our experiments did not provide any clear-cut evidence of vitamin A deficiency. An animal that is not growing well requires lesser quantities of most nutrients, and this is especially true of such fat-soluble vitamins as vitamin A. The young animals we used, having received an adequate diet prior to the time the experiment was begun, had some vitamin A stored in their livers. But, when it did begin and they grew less well, their requirement for most nutrients was reduced and the vitamin A in their livers sufficed to prevent the development of vitamin A deficiency. Studies of this kind do not necessarily identify which of the various nutrients may be most deficient under other conditions. The outcome depends upon which nutrient the animal needs during the time of the study.

Which Food Is Best

One is tempted to ask "Which is the best food for man?" Clearly, this is a nonsense question since none of the splendid foods we eat every day is wholly adequate nor is it meant to be. Therein lies the fallacy of "scientific experiments" such as ours. With the exception of the commercial dog food, all the foods fed the rats were deficient, to one degree or another, and in one or more of the nutrients required. Drawing a conclusion about the nutrient value of any one of them is like asking whether one would prefer to be deficient in vitamin A, protein, iron, or some other nutrient. Since a severe deficiency of any essential nutrient eventually causes death or serious disability, there is little purpose in making such a decision.

Milk is the only one of the foods used in the experiment that approaches nutritional adequacy when fed alone. It is designed as a sole source of food for very young animals, even though it is an inadequate source of iron. Nature overcame this inadequacy by assuring that young animals, human and non-human, born to adequately fed mothers have enough iron stored within their bodies to last until they are ready to consume other foods to supply the iron they need. Since milk does contain a rather generous supply of protein and most of the vitamins and minerals, its inclusion in the diet provides a safeguard against nutritional deficiencies. It is not an essential food, however, since necessary nutrients can be provided by many types of mixed diets.

What does one conclude about the nutritional value of spinach? Clearly, a diet consisting solely of spinach is toxic. The spinach-fed animals died sooner than they would have if they had had no food at all. Some years ago it was suggested that the Food, Drug and Cosmetic Law be amended to extend the provisions of the Delaney Clause to include "any toxic materials." In essence, the Delaney Clause says that no food containing any material known to produce cancer shall be sold. The extension of this clause would thus prohibit the sale of any food which contains toxic materials. The difficulties in this are exemplified by the feeding of spinach, which is "toxic," alone. The definition of "toxicity" depends upon the conditions of the test.

Probably few people would be willing to have spinach banned from the market. Every diet contains toxic materials (and carcinogens) and, on top of that, everything in the environment is toxic. Excessive amounts of essential nutrients are toxic, and excessive consumption of vitamins A and D has been known to cause serious disease. Excessive consumption of sugar or water, excessive oxygen in the atmosphere, excessive exposure to sunlight, excessive consumption of brown rice (the Macrobiotic Diet) are all lethal. Yet rigid prohibitions on any of these are not logical or possible.

No Absolutes

This is not to say, of course, that there are no dangers in the modern foods. In our increasingly complex society we have to use a variety of chemicals and other materials if we are to produce enough acceptable food and distribute it to the population of this country and to the world, and this clearly requires vigilance. However, there are no absolutes. Caution combined with judgment based upon experience is the only procedure that can be followed.

Finally, we should like to mention the recent controversy over cereals and bread which has had considerable exposure in the public press. The breakfast cereal that was fed in our experiments was one of the "most nutritious" cereals, according to the rating chart prepared by Mr. Choate and presented to the Subcommittee on the Consumer, Committee of Commerce, United States Senate, July 23, 1970. Therefore, it might surprise some people to learn that rats fed this material were less well nourished than those that subsisted on white bread alone. However, from what has been said, it should be clear that we feel such comparisons are meaningless.

Favorite Whipping Boy

White flour has been a favorite whipping boy of the food activists for a long time. Consider what's been said about it: John Lear writing in the *Saturday Review* emphasized that the milling of flour removed a wide variety of minerals and vitamins and quotes such values as "40 percent of the chromium, 86 percent of the manganese, 60 percent of the calcium, 78 percent of the sodium, 77 percent of the thiamin, most of the vitamin A", and so forth. Then Roger J. Williams of the University of Texas has pointed out that young rats do not grow well on bread alone, but if one adds a variety of nutrients including protein or amino acids, vitamins and minerals, they do grow well. But from this does one cry out, "Ergo! Bread is a bad food!?"

The results of this kind of an experience should surprise no one. All foods when consumed alone are inadequate and could be improved by the addition of the specific nutrients required.

What then do we do? Require that all meat be fortified with calcium and vitamin A; orange juice (which probably cannot be fortified with enough nutrients to make it nutritionally complete and still be edible) be prohibited; skimmed milk be kept off the market: This is obvious nonsense.

Roger Williams in his book, *Nutrition Against Disease*, Putnam Publishing Co., Boston, 1970, also uses another kind of illogic which should

Animals Remaining Alive in Various Groups at Different Times (six animals in each group originally)

Weeks on Diet	Spinach	Orange Juice	Skim Milk	Hamburger	Breakfast Cereal	All other Foods[1]
1	0	5	6	6	6	6
2		5	6	6	5	6
3		5	6	6	5	6
4		4	6	6	5	6
5		4	6	6	5	6
6		4	6	5	5	6
7		4	5	5	5	6
8		4	5	3	5	6
9		3	4	2	5	6
10		2	4	2	5	6
11		1	4	2	5	6
12		1	4	1	5	6
13		1	4	1	5	6

[1] Animals which received either dog food, whole milk, white bread or French fried potatoes.

be recognized. He points out that when he feeds rats a diet containing inadequate amounts of a vitamin, they do very poorly. When he doubles the amount of the vitamin in the diet, they do much better. Thus, Dr. Williams implies that if the vitamin content were quadrupled or more, they would be still better. One could as reasonably argue that since a child receiving 50,000 units of vitamin A per day will become ill but does not when he receives 5,000, he should be still healthier if he received only 500 units or, better still, 50 units. The trouble with this logic is that a child receiving only 500 units of vitamin A per day, would probably become deficient in vitamin A, and if he were receiving only 50 units a day he would certainly die.

A simpler example might be to recall that most of us need about 2,000 calories per day in order to remain healthy. Do we then conclude that if we had 3,000 or 4,000 calories per day we would be better still? Even the least sophisticated recognize that this is not true.

The addition of nutrients to foods is a well-established public health measure. Those that should be added, however, depend upon the particular food and the circumstances under which it will be consumed. The sole food of very young infants is likely to be the formula the pediatrician prescribes, and, quite obviously, it should be complete with regard to all nutrients. Such completeness may be accomplished by adding specific nutrients to the product from which it is constructed, usually a milk product. In contrast, the objective in most other fortification programs is simply to select a vehicle which will be effective in providing those nutrients which are low in a relatively large number of diets. In order to do this job, the food fortified must be one that is consumed by most people, and the fortification process must be technically feasible and relatively inexpensive. It was because bread meets those specifications that its fortification with certain B vitamins and iron was proposed many years ago. The fortification of bread or of salt with iodine, of water with fluoride, of milk with vitamin D, and of margarine with vitamin A clearly is not an attempt to make these complete foods. They have been chosen because they are effective carriers of nutrients.

Various cereal products are often reasonably satisfactory as carriers of nutrients, not only because they are widely consumed but because they are relatively inexpensive. They provide a mechanism for the delivery of nutrients to those with limited income, the population group most likely to be living on inadequate diets.

Fortification Questions

The question is often asked whether nutrients other than those now included in fortification programs should be added to bread and cereals and perhaps other foods. This question deserves continuing review. As we have indicated, the answer should be determined not by the effect observed when bread is the sole food consumed but rather by the nutritional needs of the population. If it is determined that a certain nutrient is generally low in a significant proportion of diets and if the addition of that nutrient to some food is technically feasible and economic, then it may be wise to add it. However, there are dangers in relying too heavily upon the addition of specific nutrients to foods as a primary means of protecting the nutritional quality of the food supply. The trend toward increased reliance on food fortification encouraged by the recent laws on nutrition labeling, may not be advantageous. We do stress that our knowledge of the nutritional needs of man is limited. Studies with rats and other species, if properly interpreted, are informative on that point but rarely definitive.

The consumption of a wide variety of foods, selected with some knowledge of their nutritional characteristics, is the best way to assure that a diet is safe, adequate in nutrients and low in undesirable materials. Heavy reliance upon any particular single food source should be avoided. Vigilance on the part of government, industry, and the consumer is necessary but we should not be misled by over-simplified or "scientific" experiments, with emphasis on the quotation marks.

Dr. Hegsted was Professor of Nutrition, School of Public Health, Harvard University. Dr. Ausman was Research Associate in Nutrition, also at the School of Public Health, Harvard University.
Reprinted by permission of Nutrition Today

IS THERE A FOOD SAFETY CRISIS?

E. M. Foster

Director of the University of Wisconsin's Food Research Institute delivered 14th W.O. Atwater Memorial Lecture as IFT Annual Meeting keynote address

If any of you had been living in the United States a hundred years ago—when Dr. Wilbur O. Atwater was still in the early stages of his career—you would remember a concern for food safety no less intense than the one we see today. Urbanization of America following the Civil War brought a complete change in the nation's food supply system. No longer could each family produce its own food or buy it from a nearby farmer. A food industry became necessary to supply the growing cities.

I am sure you know about some of the health problems that arose during those times. In 1869—the year Dr. Atwater received his Ph.D. degree from Yale University—*Harper's Weekly* complained that "the city people are in constant danger of buying unwholesome meat; the dealers are unscrupulous; the public uneducated." Keep in mind that this was only five years after Louis Pasteur had disproved the theory of spontaneous generation, and even before Robert Koch showed that microbes cause disease.

The milk supplies in the cities were even more hazardous than the meat. According to Otto Bettmann, "Bacteria-infected milk held lethal possibilities of which people were unaware. The root of this problem was in the dairy farms, invariably dirty, where the milk cows were improperly fed and housed.

"It was not unusual for a city administration to sell its garbage to a farmer, who promptly fed it to his cows. Or for a distillery to keep cows and feed them distillery wastes, producing what was called 'swill milk.' This particular liquid caused a scandal in the New York of 1870 when it was revealed that some of the cows cooped up for years in filthy stables were so enfeebled from tuberculosis that they had to be raised on cranes to remain 'milkable' until they died."

On top of that, "It was common knowledge to New Yorkers that their milk was diluted. The dealers were neither subtle nor timid about it; all they required was a water pump to boost two quarts of milk to a gallon."

Unscrupulous manufacturers adulterated coffee with charcoal, cocoa with sawdust, olive oil with coconut oil, butter with oleomargarine, honey with sugar, and candy with paraffin. They preserved milk with formaldehyde, meat with sulfurous acid, and butter with borax.

I think you will agree that the American people had something to complain about in those times. The result, of course, was passage of the Pure Food and Drugs Act in 1906.

Dr. Atwater died the following year. Since he lived through the events just described, I wonder what he would think of our present concerns about food safety.

This brings me to my subject, "Is There a Food Safety Crisis?" The title implies that there is one, or there may be; but I submit that the answer is no. We don't have a food safety crisis, except, perhaps, in the minds of people who are prone to worry and believe everything they hear. What we do have is a giant controversy with conflict between various factions of society over what we as a nation ought to do about food safety.

I propose here to review some of the events and circumstances that led up to this controversy. In doing this, I shall examine what we know about the actual hazards in foods as opposed to what we are asked to believe or to assume.

Historical Perspective

When I was a boy on my parents' farm in East Texas, we never worried about our food supply. All summer long, we feasted on fresh fruits, vegetables, meats, poultry and dairy products from our own garden, fields, herds, and flocks. I can remember only nine items that we bought regularly from the grocery store—flour, oatmeal, sugar, salt, baking powder, baking soda, coffee, black pepper, and occasionally a little vanilla extract. Everything else we produced ourselves.

We ate reasonably well in spring and fall as well as summer, but those Texas winters were pretty grim. Almost everything we ate was processed—by my mother, who spent all summer canning, drying, salting, and pickling food to keep us alive through the winter. Home-processed food was not exactly our first choice, but it was a lot better than going hungry. There were no such things as fresh fruits and vegetables in the wintertime, like we enjoy today.

Fortunately, we didn't know enough to ask if our food was safe. Like everyone else, we were concerned only with getting enough to eat.

I learned a little about nutritional deficiencies and foodborne disease when I attended college, but food safety still occupied a minor place in the public's consciousness. We learned about it when Congress passed the Food Additives Amendment of 1958 with its well-known Delaney clause.

Widespread concern about the safety of chemicals in our food supply increased noticeably following the appearance of Rachel Carson's book, *Silent Spring*, in 1962. It continued to grow through the 1960s and exploded in the fall of 1969, when the federal government banned cyclamates. I still remember the tension surrounding the Institute of Food Technologists' Annual Meeting in San Francisco several months later. Students from Bay Area universities picketed the meeting site, demanding an end to the use of chemicals in foods.

The decade of the '70s produced a succession of knotty problems, most of which are still unresolved. Questions have been raised about the safety of many common food ingredients, including familiar additives and even naturally occurring food components. Debate on these issues has reached the highest levels of government, including the U.S. Congress itself, where several bills that would change the Federal Food, Drug and Cosmetic Act are now being considered. Truly, food safety has become a national issue, whether it deserves to be or not.

Only a little over a year ago, for example, a two-day conference titled "Focus on Food Safety" was held in Washington, D.C. The stated purpose of the conference was to examine the safeguards that protect the American public against unsafe food—"How well do they work? Whom do they protect? How much do we need them? How safe should our food be? How safe can it be? How does one define safe?"

Featured speakers included a U.S. Senator, a Congressman, the Secretary of Agriculture, several past and present high officials of the Food and Drug Administration, top regulators from the Department of Agriculture, members of Congressional committee staffs, and an assortment of consumer activists, journalists, economists, and Washington-based attorneys.

I cite these details merely to illustrate the degree to which Americans' confidence in the safety of their food supply has been challenged. Twenty years ago, we never even thought about safety of our food. Safety was assumed. Now we question nearly everything. Had it not been for the quick action of Congress a few years ago, both nitrite and saccharin would have disappeared from the American food supply. Our nation's top regulatory agency, the Food and Drug Administration, felt legally obliged to ban both nitrite and saccharin. The agency was overruled by a body of politicans—Congress.

Needless to say, the American people are worried and confused by the mixed signals they are receiving from the many voices that speak on safety issues. Over and over again, we hear that our foods are being poisoned with unsafe preservatives, unnecessary additives, toxic pesticides, and all manner of dangerous chemicals. We hear that our diet is composed of non-nutritious junk filled with empty calories. We hear that eggs increase cholesterol levels in the blood and thereby the risk of heart attack. We also hear that cholesterol in the diet does *not* increase the risk of heart attack. We hear that diet is responsible for somewhere between 10 and 80% of all the cancer in the country, but the actual causes of cancer are not identified. We are advised to eat more polyunsaturated fats to decrease the risk of heart disease; yet we are told that polyunsaturated fats may increase the risk of cancer.

We have been warned that coffee can cause birth defects, lumps in the breast, and cancer of the pancreas; but recently we learned that coffee can reduce the incidence of breast cancer (by inducing glutathione S-transferase, which detoxifies many carcinogens). Much has been said about the cancer-promoting properties of the antioxidant butylated hydroxytoluene; but we also know that this familar additive can help prevent cancer under certain conditions.

These are just a few of the controversial safety issues that have emerged in recent years. Why do such uncertainties exist? Why has food safety become a national issue? Why are so many people worried and confused about the safety of the foods they eat?

Nature of the Hazards

Before we consider answers to questions such as these, let us identify the hazards that have been associated with food. A convenient starting place is the list prepared by FDA about 10 years ago. Named first and considered the greatest danger to consumers were foodborne toxigenic and pathogenic microorganisms. Next in order of decreasing seriousness were malnutrition, environmental contaminants, toxic natural constituents, and pesticide residues. And finally, in last place, were food additives.

To this list, I would now add reaction products that are formed during processing or preparation for eating. Nobody knows if they represent a significant hazard to man, but I have elected to list them in fifth place after naturally occurring toxicants. I reserve the right to move them to another place at some future time as more evidence accumulates.

Now, let us examine some of these hazards in greater detail:

• **Toxigenic and Pathogenic Microorganisms.** Considering the widespread concern, and in many cases near-hysteria, over the safety of foods, one might reasonably expect to find evidence of serious and widespread disease attributable to poisonous chemicals in the foods we eat; but this is not the case. Evidence of illness is limited almost entirely to incidents involving biological agents, naturally occurring toxicants, and very rarely, an inadvertent chemical contaminant from the environment.

Table 1 gives the data for foodborne disease in 1978 (published by the Centers for Disease Control in 1981). There were 154 incidents

Table 1   Confirmed Foodborne Diseases, 1978

AGENT	NO. OF OUTBREAKS	NO. OF CASES
Microbial		
Clostridium botulinum	12	58
Clostridium perfringens	9	617
Bacillus cereus	6	248
Salmonella spp.	45	1,921
Shigella spp.	4	159
Staphylococcus aureus	23	1,318
Vibrio parahaemolyticus	2	86
Other bacteria	4	59
Hepatitis virus A	5	300
Trichinella spiralis	7	35
Chemical		
Naturally occurring seafood toxins	30	96
Toxic mushrooms	1	7
Heavy metal	1	41
Other chemicals	5	19
Total	154	4,964

involving almost 5,000 cases in which the cause of illness could be established. Another 327 incidents with 5,700 cases were recognized as foodborne, but the causal agents could not be identified. These figures are typical of our experience over the past 20 years—a few hundred outbreaks each year involving a few thousand people. The vast majority of the incidents were caused by pathogenic or toxigenic bacteria.

Reporting foodborne disease to the public health authorities is not required in the U.S.; hence, the published figures must be below the true incidence. Recent estimates based on careful and conservative extrapolation from the reported figures place the actual number of cases between 1,400,000 and 3,400,000 per year. These estimates are consistent with others made over the past quarter-century.

Three points about microbial food poisoning deserve emphasis:

1. The vast majority of cases are relatively mild and of short duration. This may account for the lack of interest in investigating and reporting disease outbreaks.
2. Preventing foodborne disease is a simple matter of following good food manufacturing and handling practices. Virtually all food poisoning incidents in the U.S. are traceable to mishandling in the home or in foodservice operations. Rarely is a disease outbreak attributable to errors in commercial processing.
3. From time to time, we discover a microbiological hazard that we did not know before. In recent years, we have learned that seafood from North American coastal waters may carry *Vibrio parahaemolyticus*, just as it does in Japan, where this organism is the chief cause of food poisoning. We have discovered the cholera bacillus in seafood from the Gulf of Mexico, and we are now recognizing *Campylobacter jejuni* and *Yersinia enterocolitica* as potentially important agents of foodborne illness. This information may help us explain some of the outbreaks whose causal organisms in the past were not identified.

Not shown in Table 1 because we have no figures for illness, but potentially one of the most serious biological agents of all, are the fungal poisons called mycotoxins. The best-known and probably the most important of these is aflatoxin, which was discovered a little over 20 years ago in moldy peanut meal. Aflatoxin is a potent liver carcinogen in a broad range of animal species. Epidemiological studies in several tropical areas support the assumption that aflatoxin can cause liver cancer in man.

Controlling this hazard involves careful screening of susceptible crops including peanuts, cottonseed and corn in certain geographical areas. Insect damage to the growing crop increases the incidence and severity of aflatoxin contamination. In 1977, a high percentage of the Southeastern U.S. corn crop was found to contain aflatoxin in excess of the 20-parts-per-billion limit established by FDA. This is a recurrent problem, worse in some years than in others. In 1981, an estimated $200,000,000 worth of corn in a single Southeastern state contained violative levels of aflatoxin. Similar problems exist with cottonseed.

I should emphasize that our federal and state regulatory authorities exercise constant vigilance against the distribution of dangerous food to the public. Identifying a hazard in food that is already processed and distributed can bring disastrous financial losses to an individual processor or even an entire industry. In February 1982, for example, a fatal case of botulinal poisoning in Brussels, Belgium, led to a massive recall of canned salmon that was packed in the State of Alaska. As of late May 1982, this recall had involved more than 55,000,000 cans of salmon packed by nine different companies. Incidents such as this are rare, but they provide ample incentive to industry to produce safe products.

• **Malnutrition.** Scurvy, beriberi, rickets, goiter, pellagra, and other diseases associated with deficiency of a specific dietary component have been largely eliminated from the U.S. population. Now we hear more about problems caused by excesses of essential nutrients—sodium, selenium, iodine, and vitamin A, to name a few. The toxic level of some of these substances is surprisingly close to the optimum intake. We are urged to eat more fiber and less sugar, meat, butter, and eggs. Some experts say we get too much protein, and others say it doesn't matter. I propose to leave the malnutrition question where it is—under active debate. The issue is a popular subject of discussion, and hard facts are surprisingly scarce.

• **Environmental Contaminants.** Foods may become contaminated with potentially harmful chemicals in various ways. Table 2 lists 16 such chemicals that have been demonstrated in food at one time or another in recent years and whose safety has been questioned.

There is no doubt about the hazard from toxic heavy metals and polyhalogenated biphenyls; the problem arises in deciding what is a safe level. FDA is making a determined effort to reduce our lead intake by pressing the canning industry to eliminate or modify the traditional three-piece can.

Chlorine came under scrutiny after it was recognized that the halogen can react with organic compounds in water to yield carcinogenic products. Chlorine is the universal bactericidal agent for drinking-water treatment and is the most commonly used chemical for sanitizing food-contact surfaces. Replacing chlorine for these applications would be difficult.

Table 2   Chemical Contaminants of Concern in Food, 1970–80

Heavy metals
- Lead
- Mercury
- Cadmium
- Arsenic
- Selenium

Halogenated compounds
- Iodine
- Chlorine
- Vinyl chloride
- Trichloroethylene
- Ethylene dichloride
- Polybrominated biphenyls
- Polychlorinated biphenyls

Others
- Asbestos
- Antibiotics
- Acrylonitrile
- Diethylstilbestrol

• **Toxicants Occurring Naturally in Food.** Through the ages, man has learned that certain plants and animals are poisonous. He has managed to survive by omitting them from his diet or by treating them in some way to eliminate the poisonous factors. Table 3 lists some of the more important toxic substances that occur naturally.

Paralytic shellfish poisoning occurs on all three U.S. coasts and deserves special mention. Susceptible shellfish, usually clams or mussels, become dangerous when they ingest certain microscopic algae that carry the toxic principle. These tiny organisms, called dinoflagellates, occur normally in the sea and sometimes, for reasons that are not understood, they undergo explosive multiplication even to the point of coloring the water (hence, the term "red tide"). When this happens, the shellfish become poisonous, and harvesting must be stopped until the situation clears up. This can take months.

• **Reaction Products.** For almost 20 years, we have known that benzo(a)pyrene and other carcinogenic hydrocarbons accumulate in beefsteak during charcoal broiling. More recently we learned that nitrosamines can form in bacon during cooking, when residual nitrite reacts with secondary amines. Products dried in combustion gases are likely to have higher nitrosamine levels than those dried by indirect heating. The discovery of mutagenic substances in charred beef and fish led to similar findings with grilled hamburger and other heated

Table 3 A Sampling of Toxicants Naturally Occurring in Foods

Plant Sources
Cyanide
Safrole
Cycasin
Solanine
Quercetin
Oxalates
Bracken fern
Goitrogens
Hemagglutinins
Toxic mushrooms
Pyrrolizidine alkaloids

Animal Sources
Ciguatoxin
Tetrodotoxin
Paralytic shellfish poison

foods. Efforts are now underway to identify the mutagenic compounds and determine if they have any real significance to human health.

• **Pesticide Residues.** There is little evidence of danger from residues of pesticides in food. These agents are closely regulated, and the experts are satisfied that they pose little hazard to consumers.

• **Food Additives.** The most heated controversy centers around the food additives. Table 4 lists 28 substances whose safety has been seriously and openly challenged in the 12 years since cyclamates were removed from the American food supply.

Six items have actually been banned from use in food during that time: Red No. 2, Violet No. 1, carbon black, salts of cobalt, cyclamates, and diethylpyrocarbonate. All other items in Table 4 still are being used, some in the midst of continuing controversy and debate. Nitrite and saccharin are temporarily protected by a Congressional moratorium that forbids banning until further information is available.

Overall Assessment

The real problems, the ones we know about and have a valid basis to fear, are the biological agents—pathogenic and toxigenic bacteria, foodborne viruses and parasites, fungal toxins, and toxic natural components of edible plants and animals. We know how to prevent all of these.

Everything else is either hypothetical or rare—accidental and pre-

Table 4 Some Food Additives Whose Safety has been Questioned in the U.S. since 1969

Salt	Sulfite
	Nitrite
Sugar	Nitrate
Xylitol	Phosphate
Mannitol	Cobalt salts[a]
Aspartame	BHA, BHT
Saccharin	Caramel
Cyclamates[a]	Caffeine
	Carrageenan
Red No. 2[a]	Monosodium glutamate
Red No. 3	Diethylpyrocarbonate[a]
Red No. 40	Modified starches
Yellow No. 5	Hydrogenated fats
Violet No. 1[a]	Brominated vegetable oil
Carbon black[a]	Synthetic colors and flavors

[a] Ultimately banned

ventable with reasonable care, just as auto accidents are. There is absolutely no evidence that we are using unsafe food additives, yet our concerns about foodborne hazards seem to be inversely proportional to how much we know about them. The less we know, the more we worry.

Many Americans are confused, scared, and concerned about food safety because they don't know what to believe or whom to believe. For the past 10 years, we have been exposed to claims by presumably reputable and knowledgeable people that one element or another of our diet is unsafe. Other presumed experts say that is not true. What is the layman to do in the face of such obvious contradiction?

Recently, a dentist advised a young woman friend of my wife's to rinse her mouth with fluoride solution to treat an oral disorder, but the patient didn't think she wanted to do it. "Too many young people are dying of cancer these days," she said, obviously convinced that fluoride is carcinogenic. Apparently, she didn't know that her city's water supply is fluoridated.

It is difficult to understand how we got into this state of affairs until we recognize the unique, almost mystical properties of food. Food has the largest burden of symbolic, ceremonial, and religious overtones of any component of our daily lives. Food is something special. We feel comfortable with food and nervous without it. Concern about our food supply can make us irrational. We are seeing today.

Food scientists have been modifying, manipulating, and improving our food since the time of Nicolas Appert. Wars accentuate the need for preserved food, and World War II in particular provided an environment for technological innovation that carried over into the post-war civilian food supply. The late 1940s and the decade of the '50s were the heyday for food technologists interested in developing new goodies to please the American palate and ease the housewife's job of food preparation.

Then came the consumer movement—which might better be called the anti-industry movement—of the 1960s, with a ready-made opportunity to attack the food industry for adding, in the words of a well-known former U.S. Senator, "all those unsafe, untested, unnecessary chemicals to our food supply."

The consumer movement was made up of a loose federation of activists from many and varied backgrounds who shared antipathy for the food industry. They had a simple and effective mode of action. First, they would seize the high moral ground as defenders of the individual consumer against his monolithic foe, the food industry. From this vantage point the activists would attack industry for manufacturing and selling unsafe and unwholesome food. Anyone who did not agree with the activists was dismissed and discredited as a toady of industry.

Coincident with their attacks on industry, the activists mounted a campaign to change people's food habits. They urged us to eat less processed food, avoid food additives, and cut down on animal products. They recommended that we eat a "natural" diet with more whole-grain cereals and other unrefined foods obtained from plants.

How successful have the activists been? I have no specific figures, but news reports and observations tell me that their efforts have not been in vain. In every group, one finds people who have changed their food habits. They are cutting down on salt, or they have quit eating bacon, or they have given up beef, or they eat only one egg per week, or they avoid sweets. Far more people now are reading labels, looking for the name of additives they want to avoid. The phenomenal growth of the so-called health-food industry attests to the magic of the word "natural" which nobody can define but everybody, including beer advertisers, uses to lure a gullible public. Natural is supposed to mean pure and safe, neither of which it does. Recently, my wife had a conversation with a young professional woman who was firmly opposed to what she called "processed and synthetic foods." For her, everything had to be "natural." They got to talking about nitrite in cured meats, and my wife mentioned the alternative possibility of botulism. "Oh, that would be all right," said the young woman. "Botulism is natural."

How did industry react to its attackers? I would characterize the response as very quiet. With a few notable exceptions, industry went about its business without public utterance. There was occasional wringing of hands in board rooms and plenty of complaining and worrying in private, but that was as far as it went. In typical marketer fashion, most companies determined to give the customer what he wants. If he wants "no additives" and "everything natural," that's what he will get. Meanwhile, the same companies usually sell other products containing a full quota of chemical additives. The hypocrisy doesn't seem to bother them. This is called opportunism in the marketplace.

In retrospect, it is not difficult to see why the activists have been so successful in elevating food safety to a national issue. It is because they had no opposition of consequence and because we are so ignorant. As a modern philosopher once said, one cannot counter anecdotal data with no data. Claims that something *might be* unsafe cannot be laid to rest by a simple retort that it *might not be* unsafe. As a friend of mine said recently, we don't know enough to solve the food safety problems, but we know too much to ignore them.

Well, what can we do about the controversy over food safety? I have four suggestions:

1. The first thing is to get our priorities straight. Let's put our efforts on the *real* hazards in life and quit dissipating our energies on hypothetical and imaginary dangers. Table 5 gives some of the leading causes of death in the U.S. as taken from an article by Dr. Arthur Upton in the February 1982 issue of *Scientific American*. The figures tell us that smoking kills 150,000 people every year. That is almost three times the number of Americans killed during the entire Viet Nam War. Alcohol kills 100,000 people every year; motor vehicles kill 50,000. Not shown is the fact that drunk driving kills an average of 70 people *every day*. These are real numbers, and these are real dead people.

    Look at the bottom of the list. Not a single fatality attributable to the much-criticized food constituents, pesticides, antibiotics, and spray cans.

2. The second thing we might do is use a little common sense. At the height of the nitrite controversy, critics of the food supply were demanding that nitrite be banned from cured meats on the grounds that it may cause cancer. They were willing to jeopardize a $12-billion industry and allow the consumer to take the risk of botulism—knowing full well that the ban would reduce his exposure to nitrite by less than 10%. That doesn't make much sense to me.

    While we are talking about common sense, I hope the present urge to reduce salt intake is not carried to extreme in cured

Table 5 No. of Deaths each Year in the U.S. Attributable to Various Causes

CAUSE	NO. OF DEATHS/YR
Smoking	150,000
Alcohol	100,000
Motor vehicles	50,000
Handguns	17,000
Electric power	14,000
Motorcycles	3,000
Swimming	3,000
Surgery	2,800
X-rays	2,300
Railroads	1,950
General aviation	1,300
Construction	1,000
Bicycles	1,000
Hunting	800
Home appliances	200
Fire fighting	195
Police work	160
Contraceptives	150
Commercial aviation	130
Nuclear power	100
Mountain climbing	30
Power mowers	24
Scholastic football	23
Skiing	18
Vaccinations	10
Food coloring	0
Food preservatives	0
Pesticides	0
Antibiotics	0
Spray cans	0

meats. Though both are active, salt is much more important than nitrite in protecting cured meats against botulinal toxin development.

3. The third and by far the most difficult thing we need to do is identify the real dangers in our food supply and learn how to control them. Unfortunately, we do not have the basic information required to do this, and a great deal more research will be necessary before we can reach valid conclusions. Calling for more research is a typical reaction from a scientist, but that is the only way to get the facts we must have. The alternative is to continue

as we are. Who is going to do the research and who is going to pay for it are two questions that I cannot answer. Recently, FDA Commissioner Arthur Hayes urged the food industry to step up its research efforts in food safety. I wish him well in this approach, but I am not sanguine about his chances.

4. My fourth suggestion is to develop a mechanism for getting the facts about food safety to the public. Up to now, our people have been subjected to a great deal of misinformation and supposition that was based not on fact but simply on the biases of the communicator.

I would close with the observation that things are not nearly as bad as they look. We have very little evidence of hazards in our food supply, and the ones we know about we can control. In spite of all the claims about carcinogens in food, the death rates from cancer for all organs except lungs are declining or remaining level. Life expectancy is increasing. The average American now lives 21 years longer than he did when I was born. That is an entire generation gained in one lifetime.

We must be doing something right.

(Reprinted from *Food Technology* 1982, vol 36 (8):92 by permission of Institute of Food Technologists)

Appendix III
List of Commonly Used Food Additives and Explanation of Their Functions

A.3.1 SOURCE

The following list and accompanying notes have been extracted from a Government of Canada publication "Food Additive Pocket Dictionary" (Cat. No. H49–10/1980 E Rev., published by Authority of the Minister of National Health and Welfare, Educational Services, Health Protection Branch, Ottawa, Ontario, K1A 1B7). Only the most commonly used substances approved for food use in Canada are included in this abbreviated list.

A.3.2 CATEGORIES OF ADDITIVES AND THEIR FUNCTIONS

Additives listed in Table A.3.1 in alphabetical order are identified by a code which explains their specific function in a food. Some substances may be used for more than one function. The codes and the explanation of an additive's function are shown in Table A.3.2. To find why a given substance has been used in a given food, Table A.3.1 should be consulted first to find the code for the additive (if listed); the general function of the additive is then found in Table A.3.2. Only the most common additives found often on the labels of consumer-packaged foods are listed. This list is provided as an example of functions of the additives used and not as an exhaustive listing of additives permitted in Canada or GRAS substances used in the U.S.A.

Table A.3.1 List of Common Food Additives Used in Processed Foods

ADDITIVE	CODE
Acacia Gum (Gum Arabic)	Tm,GP
Acetic Acid	pH,P
Esters of Mono and Di-Glycerides	Tm
Aluminum Sulphate	F,Sm,X
Amylase	Fe
Annatto	C
Anthocyanin	C
Ascorbic Acid (Vitamin C)	BM,P
Aspergillus Oryzae Enzyme	BM
Benzoic Acid	P
Bromelain	Fe
Butylated Hydroxyanisole (B.H.A.)	P
Butylated Hydroxytoluene (B.H.T.)	P
Caffeine	X
Calcium Aluminum Silicate	Ac
Calcium Carrageenan	Tm
Calcium Citrate	Tm,pH,F,S,Yf
Calcium Phosphate Dibasic	Tm,F,X,pH,Yf
Calcium Propionate	P
Calcium Silicate	Ac,X
Caramel	C
Carbon Dioxide	Pd
Carob Bean Gum (Locust Bean Gum)	Tm
Carotene	C
Carrageenan	Tm
Cellulose, Microcrystalline	Tm,X
Chlorine Dioxide	BM
Citric Acid	P,S,pH,X
Ethylenediaminetetraacetate (EDTA)	S
Ethylene Dichloride (1,2-Dichloroethane)	CE
Ethylene Oxide	X
Gelatin	Tm
Glucono Delta Lactone	pH,X
Glycerol	H,CE,GP
Guar Gum	Tm
Hexane	CE
Hydrochloric Acid	pH,Sm
Iso-Ascorbic Acid	P
Isopropyl Alcohol	CE
Karaya Gum	Tm
Lactic acid	pH
Lecithin	Tm,R,P
Lipase	Fe
Magnesium Aluminum Silicate	X

Table A.3.1 (Continued)

ADDITIVE	CODE
Magnesium Silicate	Ac,GP,X
Malic Acid	pH
Mono and Diglycerides	Tm,Af,H,R,CE
Nitrogen	Pd
Nitrous Oxide	Pd
Papain	Fe
Pectin	Tm
Pepsin	Fe
Phosphoric Acid	pH,S,Yf
Polyethylene glycol	Af
Polysorbate	Tm
Potassium Acid Tartrate (Potassium Bitartrate)	pH
Potassium Aluminum Sulphate	F,pH,X
Potassium Bromate	BM
Potassium Chloride	pH,Yf,Tm
Propionic Acid	P
Propyl Gallate	P
1,2-Propylene Glycol	Ac,H,CE
Rennet	Fe
Riboflavin (Vitamin B-2)	C
Sodium Acid Pyrophosphate	Tm,pH,S
Sodium Alginate	Tm
Sodium Aluminum Phosphate	pH,Tm
Sodium Aluminum Silicate	Ac
Sodium Benzoate	P
Sodium Bicarbonate	pH,Sm,X
Sodium Citrate	Tm,S,X,pH
Sodium Hydroxide	pH,Sm
Sodium Lauryl Sulphate	W
Sodium Nitrite	P
Sodium Phosphate Dibasic	Tm,pH,S,X
Sodium Silicate	X
Sodium Stearoyl-2-Lactylate	BM,Tm,W,X
Sodium Tripolyphosphate	S,pH,Sm,X
Sorbic Acid	P
Sorbitan Monostearate	Tm
Sorbitol	R,H,X
Tartaric Acid	P,pH
Tocopherols	P
Wood Smoke	P
Xanthan Gum	Tm
Xylitol	X
Zinc Sulphate	Yf

Table A.3.2 Explanation of Functions of Various Food Additives by Categories

CODE	CATEGORY OF ADDITIVES	FUNCTION IN FOODS
Ac	Anticaking Agents	Keep powders (for example, salt) free-running.
Af	Antifoaming Agents	Prevent undesirable foaming during the manufacture of certain foods such as in the making of some jams.
BM	Bleaching, Maturing and Dough-Conditioning Agents	Act on flour to give a product of consistent quality and color.
C	Coloring Agent	Give foods an appetizing appearance. Factors such as processing, storage and seasonal variation can result in unattractive or unfamiliar color. Usually, the word "color" appears on a label, not the specific chemical or common name.
CE	Carriers or Extraction Solvents	Act as vehicles for various food components, either to keep them in the food or to remove them from the food; for example propylene glycol is used to dissolve color used in some margarines.
F	Firming Agents	Maintain the texture of various foods, such as canned tomatoes.
Fe	Food Enzymes	Promote desirable chemical reactions in food. Rennet, for example, is an enzyme used to curdle milk in cheese making.
GP	Glazing and Polishing Agents	Make food surfaces shiny and in some cases offer protection from spoiling. They are used mainly in candies.
H	Humectants	Keep foods moist as in shredded coconut and marshmallows.

Table A.3.2 (Continued)

CODE	CATEGORY OF ADDITIVES	FUNCTION IN FOODS
Ns	Non-nutritive Sweeteners	Sweeten food without adding calories or other food value to the food.
P	Preservatives	Are used to prevent or delay undesirable spoilage in food, caused by microbial growth or enzymatic and chemical actions. Antimicrobial agents prevent the growth of molds, yeast or bacteria in foods. Antioxidants slow down the process of fats turning rancid and frozen fruits turning brown.
Pd	Pressure-Dispensing Agents	Act as propellants to dispense foods such as whipped toppings from aerosol containers.
pH	pH-Adjusting Agents	Reduce or increase the acidity (sourness) of food. Some are also components of leavening agents and help to make baked products light and fluffy.
R	Release Agents	Help food separate from surfaces during or after manufacturing. Mineral oil, for example, is applied to baking pans and facilitates the removal of baked goods without sticking or crumbling.
S	Sequestering Agents	Combine with metallic elements in food, thereby preventing their taking part in reactions leading to color or flavor deterioration. For example, the addition of a sequestrant to canned lima beans prevents darkening of the product because the ions from iron and other trace

Table A.3.2 (Continued)

CODE	CATEGORY OF ADDITIVES	FUNCTION IN FOODS
		metals in the canning water are bound by the additive and consequently are unavailable for other reactions.
Sm	Starch-Modifying Agents	Alter the property of starch in order to withstand heat processing and freezing and thus maintain the appearance and mouth-feel of foods.
Tm	Texture-Modifying Agents	Contribute or maintain desirable consistency in foods.
W	Whipping Agents	Assist in the production and maintenance of stable whipped products.
X	Miscellaneous Agents	Include a variety of other food additives, such as carbonating agents in soft drinks.
Yf	Yeast Foods	Are substances that serve as nutrients for yeasts such as those used in the manufacture of beer and in the making of bread.

Index

Acceptable Daily Intake, 209
Additives, food. *See* Food additives
Agar, 40
Agglomeration, in spray drying, 227
Aging:
    of beef, 156–157
    of wine, 187
Agricultural chemicals, 212
Agricultural sciences, and food processing, 2
Agricultural terminology, 3t
Aflatoxin, 291–292
Alcohol, energy content of, 15t, 196
Alcohol fermentation, 176–178
Alcoholic products, sources for, 177t
Aluminum, in food packaging, 255, 261
Amino acids, 14–17
    essential, 17t
Ammonia, in refrigeration, 235
Animal fats, saturation of, 112t, 172
Annato cheese color, 206
Antibiotics:
    as food additives, 211
    in milk, 129
Anticaking agents, as food additives, 304
Antifoaming agents, as food additives, 304
Appert, Nicholas, 236

Apples:
    as a source of pectin, 84
    composition of, 8t
    controlled atmosphere storage of, 76t
    in jams and jellies, 83t, 84
Aquaculture, 171
Ash, analysis for, 5
Aspartame, 194

Baby Duck, 187
Bacillus cereus, 39, 290t
Bacteria, 30–31
Baking of bread, 99–105. *See also* Bread
Baking powder, 105
Barley, malting of, 107–108
Beef. *See* Meat
Beer, 178–185
    ale, 184
    components in brewing, 178
    draught, 184
    fermentation of, 183–184
    hops extraction 181
    lager, 183
    wort, boiling of, 181
Benzoic acid, as preservative, 211

NOTE: A page number followed by **t** indicates a table.

308 Index

Benzoates:
    in soft drinks, 195
    in marmalades, 211
Beverages, consumption of, 176t
BHA, 205t, 295t
Binders, in meat products. *See* Meat
Biological Oxygen Demand (BOD), 55. *See also* Waste treatment
Biotechnology, 45
Birdseye, Clarence, 230
Blanching, 78–81. *See also* Fruits and vegetables
Bleaching agents, as food additives, 304
Blow molding, in packaging, 256
BOD, 55
Botulism, 38, 39t, 292, 296
    in canned foods, 237
        protection against, by sodium nitrite, 165–166
Bran, wheat. *See* Wheat
Bread, 99–105
    composite flours, 103
    composition of, 103
    CO_2 in dough, 101
    crumb, texture, 103
    crust formation, 102
    dough, fermentation of, 101
    gluten, 101
    ingredients, 100–101
    loaf volume, 103
    proofing, 101
    sour dough, 101
    sponge dough method, 101
    staling of, 104
    starch degradation in storage, 104
    straight dough method, 101
Breakfast cereals, 106–107
    extrusion cooking, 107
Breeding, 2
Bregott, 144
Brewers' yeasts, 45, 183
Brewing of beer, 178–185. *See also* Beer
Brix, degree of, 69, 85
Butter, 141–145
    blends, 144
    composition of, 117t
    flavor, from diacetyl, 45, 133, 143

Butterfat:
    in ice cream, 145
    saturation of, 112t, 144
Buttermilk, 134

Caffeine:
    in drugs, 196
    in soft drinks, 194
    in tea or coffee, 192
Cakes, 105
Calcium:
    in cheese, 137, 140t
    in ice cream, 145
    and osteoporosis, 149
    and oxalic acid, 89
Camembert, 139t, 140. *See also* Cheese
Campylobacter jejuni, 39, 291
Canada Food Guide, 20–21
Cancer, and diet, 289
Candling, eggs, 203
Canning, 236–243. *See also* Heat preservation
    aseptic, 239t,
Canola, 111. *See also* Oilseeds
Caramel, in beer, 184t
Carbohydrates:
    components of, 5
    energy content, 15t, 120
    in flour, 97
    in foods, 8t, 9t
    in photosynthesis, 73
Cardiovascular disease:
    and cholesterol, 131, 289
Casein. *See* Milk protein
Cellophane, in food packaging, 256t, 259
Cereals, 93–109
    breakfast, 106
    dietary fiber, 5, 96, 121
    malting of barley, 107–108
    nutritional importance, 119–121
    wheat kernel structure, 94
Cheddar, 136, 140. *See also* Cheese
Cheddaring, 139
Cheese, 136–141
    calcium retention, 137, 138, 140t
    cheddar, 136, 137t, 140t
    cheddaring, 139

Cheese (*continued*)
    composition of, 140t
    processed, 141
    rennet, clotting, 138
    ripening, 140–141
    starter culture, 43, 138
Chicken. *See* poultry
Cholesterol, and heart disease, 121, 289
Churning, butter, 143
Citric acid, 84, 195
Clostridium botulinum:
    in canned foods, 237
    control by nitrite, 165
    food-borne illness, 38, 290t
Cobalt 60, 243
Codex Alimentarius, 244
Coffee, 189–192
    beans, processing of, 190
    caffeine content, 192
    flavor of, 190
    health hazards, 289
    instant, 191
Cold pressing. *See* oilseeds, expeller pressing of
Coliforms, 39
Collagen, in meat, 155t, 160
Color:
    additives, 304
    annato, 206
    in hyperactivity, 216
    nutritional significance, 216
    of red meats, 160
    Red No. 2, 208
Commercial sterility, 237
Concentration:
    processing of milk by, 146
    of solutions, definitions, 66–69
Controlled atmosphere storage, 76
Cookies, 105
Corn, 106
Corned beef, 165
Coronary heart disease, 120–121, 131, 289
Cottage cheese, 134–135, 136, 137
CO_2:
    in bread dough, 101
    in storage of fruits and vegetables, 75–77

    in food packaging, 263–264
    soft drinks, carbonation of, 195
    solubility, 263
Coxiella burnettii, 38, 130
Cream, dairy, fat content of, 132–133
Cryogenic freezing, 236
Crystallization of sugar, 86–89
    molasses, 87
    seed crystals, 87
    supersaturated solution, 86–87
Cultured dairy products. *See* Milk
Cultures, microbial. *See* Microbial cultures, Fermentation
Curing of meat. *See* Meat
Cyclamates, 208, 294, 295

Dairy cattle, 2
Dairy products and technology, 125–149. *See also* Milk
Davis, Adelle, 149
Deboning of meat:
    hot, 158
    mechanical, 159
    nutrient losses, 24
Dehydration of food. *See* Food dehydration
Delaney Clause, 207, 283
Diacetyl. *See* butter
Diet, and cancer, 289
Dietary fiber, 5, 17–18, 96, 121
    in cereals, 121
    recommended intakes, 18
Direct heating, in UHT process, 240. *See also* UHT processing
Diverticulosis, 18, 121
Dough formation, in bread, 100–101. *See also* Bread
Drying of food. *See* Food dehydration
Dumplings, 105
Durum wheat. *See* Wheat

Efficacy, of food additives, 209
Eggs, fresh, 203
Eggs, dried:
    whole, 203
    whites, functional properties of, 203
    yolks, in mayonnaise, 203

310 Index

Emulsifiers, 118
   in ice cream, 146
   in mayonnaise, 203
   in processed cheese, 141
Emulsions, 117–119
   meat, 169
   stability, 118
   Stokes' Law, 118
Endosperm, 96, 97
Energy:
   balance, calculation of, 66, 273–274
   content of foods, 13–15
   from industrial fuels, 55
   from methane, 43–44
   requirements of humans, 14–15
   units of measurement, definitions, 15t
Enriched flour. See flour, enriched
Enzymatic browning, 81
Enzymatic reactions:
   in fish, 171
   in plant materials, 81t
Enzymes:
   as food additives, 304
Enzymes, in foods:
   lactase, 149
   lipase, 81, 129
   maltase, 81t
   peroxidase, 80
   phosphatase, 129, 130
   rennin, rennet, 135, 138
Escherichia coli, 39
Essential nutrients, 13
Ethylene gas, for ripening of fruit, 74
Eutectic point, in freezing, 231–232
Extrusion:
   cooking of cereals, 107
   of food packaging materials, 256, 259, 260

Fat:
   analysis for, 6t
   animal, saturation of, 172
   content in foods, 7, 8t, 9t, 127, 140t, 145t, 164t
   globules in milk, 127
   oxidation in dry foods, 229
   saturated and unsaturated, from various sources, 112t
   See also Oils, Oilseeds, Butter
Fatty acids:
   cis and trans, 116
   melting point of, 112
   saturation of, 112
   structure, 111
FDA, 62, 64t
Feingold, D. B., 216
Fermentation:
   alcoholic, 176–178
   microbial, for food, 42–45
   single cell protein, 44–45
   of tea, 190
   See also Beer, Bread, Milk, Wine
Fiber. See Dietary fiber
Films, packaging. See Packaging materials
Firming agents, as food additives, 304
Fish, 170–172
   by-catch, 170
   consumption, 172t
   liver oil, as vitamin D source, 172
   oil, saturation of, 112t, 172
Flexible packaging. See Food packaging, Packaging materials
Flour, 97–99
   bleaching and maturation, 99
   in bread-baking, 99–101. See also Bread
   composite, 103
   enriched, 97
   extraction number, 97
   middlings, 99
   milling process, 98–99
   in pastry, 105
Fluidized bed:
   drying, 224t
   freezing, 234t
Foam, whey protein, 204, 205
Food:
   chemical components of, 4
   colors, as additives, 304
   contaminants, 212–213, 292
   delivery chain, 51
   Guide, Canada, 20–21
   and health hazards, 289–294
   infection, 38
   intoxication, 38

Food (*continued*)
  liquids, component concentration, 66–68
  nutrients, effects of processing, 22–23, 246
  poisoning, 38–40
  preservatives, as food additives, 305
  processing, science elements of, 52
  product development, 59
  proximate composition, 4
  raw materials, 3
  storage, 56, 72, 93, 128, 146–147, 241–242
  water activity in, 36, 37t
Food additives, 199–219
  Acceptable Daily Intake, 209
  antibiotics, 211
  categories of, 205t, 304–306
  in cheese spread, 200t
  chemical preservatives, 208, 210–212
  cyclamates, 208, 294, 295t
  definition (Canada), 206
  definition (USA), 207
  Delaney Clause, 207, 283
  efficacy data, 209t
  functions, 301–306
  government regulations, 206–208
  GRAS list, 207
  and hyperactivity, 216
  labelling requirements, 213–214
  risk-benefit evaluation, 208, 209, 210t
  saccharin, 208
  safety, 215, 294–295, 298t
  testing procedures, 208–210
Food-borne illness, 38–40
Food components:
  carbohydrates, 5, 13
  dry matter, 5
  fats, 5, 111–112
  minerals, 5
  non-fat solids, 5
  protein, 5, 14, 16–17
  proximate, 4–5
  tests for, 6t
  total solids, 5
  water, 5–7
Food dehydration, 223–229
  agglomeration, 227
  flavor defects, 225
  freeze-drying, 225
  heat effects, 225
  heat of vaporization, water, 224
  methods, 224t
  packaging, 229
  quality defects, 224–225
  spray drying, 225–227
  technological principles, 223–224
Food energy:
  human requirements, 14–15
  sources, 13–14
Food freezing, 230–236. *See also* Freezing, principles of
  cryogenic process, 236
  eutectic point, 231–232
  freezer burn, 233
  mechanical refrigeration, 235
  nutrient retention, 236
  quality defects in storage, 233
  technological principles, 234t
  texture damage, 230
  water behavior, 230, 231
Food industry, 49–65
  components of, 49–59
  government regulation, 62–65
  plant organization, 53–59
  primary, secondary, tertiary, 49–50
  processing lines, 53–54
  quality control, 59–62
  support services, 53–54
Food irradiation, 243–246
  permitted uses, 244
  preservation principle, 243
  radappertization, 243
  radurization, 243
  regulation of, 244, 246
  safety, 244–245
Food packaging, 249–265. *See also* Packaging materials
  cost, 264
  CO_2, 263–264
    solubility in foods, 263, 264
  dried foods, 229, 262
  flexible pouches, liquids, 257
  form-seal principle, 257
  hygroscopic foods, 262
  inert gases, 229, 262

Index   311

Food packaging (*continued*)
    in marketing, 249
    materials, properties of, 250–259. *See also* Packaging materials
    modified atmosphere, 263, 264
    purposes, 249–250
    shrink-wrapping, 261
Food preservation, 221–248
    by chemical additives, 210–212
    by drying, 223–229. *See also* Food dehydration
    effects on macronutrients, 246
    by freezing, 230–236. *See also* Food freezing
    by heat, 236–243. *See also* Heat preservation
    by pH control, 243
    principles of, 221–222
    by water activity control, 243
Food spoilage:
    canned foods, 237
    chemical, causes of, 223
    microbial, 31–34
    types of, 221
Form-seal, in food packaging, 257
Freeze-drying, 225
Freon, in refrigeration, 235
Freezing:
    of food, 230–236. *See also* Food freezing
    principles of, 230, 231
    of water, phase diagram, 231
Freezing point:
    depression, 231
    of foods, 233t
Fruit:
    botanical characteristics, 71
    juice products, 82–83
    pectin, content in, 84
Fruits and vegetables, 71–92
    blanching, 78–79
        peroxidase indicator, 80
    characteristics, 71
    composition of, 8t, 71, 72t
    controlled atmosphere storage, 76
    enzymes in processing of, 79–80
    ethylene, in ripening of, 74
    mechanical harvesting, 3, 74

    nutritional value, 71, 88–90
    oxalic acid in, 89
    peeling, 77
    photosynthesis, 73
    preparative processes, 77
    prevention of browning, 81
    respiration, 73, 75–76
        anaerobic, 75
    storage of, 72, 75–77
    vitamin C content, 82, 88–89
Fructose, in soft drinks, 194
Functional ingredients, 201–205
    in ice cream, 201
    of microbial origin, 45
Functional properties:
    of food ingredients, 200–203
    of packaging materials, 251–259
Fumigants, 212t
Fungicides, 212t

Galactose, 128t, 149
Gamma-rays, 243
Gas, inert, in packaging, 229, 262
Gas permeability, packaging materials, 262
Gay-Lussac, 177
Gel, pectin, 84
Germ, wheat. *See* Wheat
Glass, in food packaging, 252–253
Glassine, paper, 253t
Glucose, 73, 128t, 149
Gluten, 101
Glycerol:
    in fat, 111
    as humectant, 217t
Glycogen:
    in fish, 171
    in meat, 156
Glycolysis, 156
Government regulations, 62–65
GRAS list, 207
Growth, microbial, 31–34

Harvesting, mechanical, 3, 74
Health foods, 24–25, 296
Health hazards, of foods, 289–294
Health Protection Branch, Canada, 62
Heart disease. *See* Cardiovascular disease

Heat:
    exchanger, 130, 131
    of fusion, 234
    latent, 274
    in respiration of fruits and vegetables, 73, 75–76
    specific, 274
    of vaporization, 224
Heat preservation:
    acid foods, 237–238
    botulism, 237, 292, 296
    canned foods, shelf life, 236
    commercial sterility, 237
    hydrogen swell, in cans, 254
    low acid foods, 238
    pouch-pack, 239, 240
    principles of, 236–237, 239t
    retorts, 238–239
    sanitary can, 238, 255t
    Tetra-pak packaging, 240
Heat sterilization of food, 236–243. *See also* Heat preservation
Heeney, Bill, 230
Hemoglobin, 160
Herbicides, 212t
Hexane, in oil extraction, 113
Home-processed foods, 215, 237, 288
Homogenization, of milk, 132
Hops, in beer, 181
HTST pasteurizer. *See* Milk
Humectants, 37, 216–217
    as food additives, 304
Human milk, composition of, 8t, 149
Hydrogenation, of vegetable oils, 116
Hydrogen peroxide, in UHT processing, 240, 262
Hydrogen swell, 254
Hydrostatic sterilizer, 239t
Hyperactivity, and food additives, 216

Ice cream, 145–146
    artificial flavors in, 146
    composition of, 145t
    functional ingredients, 201–202
    overrun, 145
    texture defects, sandiness, 146
Ice crystals, in food freezing, 232–233

Income, disposable, for food, 51t
Index, refractive, 82, 84t
Ingredients in foods, 202–205
    functional properties, 201–202
Injection molding, in food packaging, 256
Instantization, 227
Interesterification, of fats, 116
Intermediate moisture foods, 216
Iodine:
    in fish, 173
    and goiter, 14t
    recommended intake, 19t
Iron:
    and anemia, 14t, 281
    in meat, 172–173
    recommended intake, 19t
Irradiation of food, 243–246. *See also* Food irradiation

Jams and jellies, 83–85
    composition of, 9t, 83
    pectin, 84
    refractive index of sugar, 82, 84t
    specifications for, 83
    technology, 84
Junk foods, 107, 195

Kamaboko, 171
Kefir, 134
Kilning, in malt production, 109
Kraft paper, 253t
Kwashiorkor, 14t
K-2 blancher, 79–80

Labelling:
    ingredient, 213, 214t
    nutrition, 214, 215t
Lactalbumin, in milk, 5, 128t
Lactase, enzyme, 149
Lactic acid:
    in dairy products, fermented, 133–134, 138
    in meat, 156
Lactobacillus, in fermented dairy products, 134, 135t, 139t

Lactobacillus acidophilus, 134, 135t
Lactoglobulin, in milk, 5, 128t
Lactose. *See* Milk
Latent heat, 274
    of fusion, 234
    of vaporization, 224
Lager beer, 183. *See also* Beer
Lamination, in food packaging, 259–262. *See also* Packaging materials
Leavening, by baking powder, 105
Lecithin, 118, 120
Leeuwenhoek, Antonie van, 29
Legumes, 93
Leuconostoc citrovorum, 135t
Limburger cheese, 137t. *See also* Cheese
Linoleic acid:
    as essential nutrient, 121
    in fats, 112t
    structure, 111
Lipase, enzyme, 81, 129
Lipids, 111–112. *See also* Fats, Oils
Liqueurs, 189
Liquid nitrogen, for freezing, 236
Liquor, 177t, 188–189
    Control Boards, provincial, Canada, 64t
    government regulation, 64t
    Treasury, Department of, (BATF), USA, 64t
Low acid foods, 238
Lye, in peeling fruits and vegetables, 77
Lysine, structure of, 17t

Macaroni, other pasta products, 105
Macronutrients, 7–12
Magnesium, as micronutrient, 19t
Maillard, browning reaction, 147, 221
Malt:
    in beer-making, 178
    processing of, 107–108
Maltase, enzyme, 81t
Maltose, 101, 109
Margarine, 114–117
Mashing, in beer-making, 178
Mass balance, 65–67, 267–273
    basis, for calculation, 267
    general method, 269
    for inventory control, 67
    Pearson's square, 271–273
Material balance. *See* Mass balance
Maturity, for harvesting, 3
Mayonnaise, 117, 203
Meat:
    aging, 156–157
        electrical stimulation, 157
    by-products, 154
    collagen, in tenderization, 160
    composition of, 8t, 154–155
    cooking, 160
        in sausage manufacture, 168
    curing, 162–166
        chemical reaction, 165
        and nitrosamines, 165
    fat, saturation of, 112t, 172
    fresh, technology of, 158–162
    glycogen, in rigor mortis, 155, 156
    glycolysis, in living muscle, 156
    grading, 160, 161t
    hot deboning, 159
    as iron source, 172
    least cost formulation, 168
    mechanical separation ("deboning"), 159
    muscle, 155–156
    myoglobin, 160
        in curing, 165
    packaging, oxygen permeability, 262
    pH of, 156
    production, efficiency of, 153, 154t
    protein, 155
        collagen, 155, 160
        in diet, 172
        nutritional quality, beef, 18t
    in rat-feeding experiments, 281
    rigor mortis, 155–157
    sausage products, 166–169
        composition of, 9t, 167
        fillers and binders, 166–167
        smoking, 169
    sources of, 153–154
    tenderization of, 157, 160
Mechanical harvesting, 3, 74
Mellorine, 203
Methyl alcohol, in home brews, 215

Metmyoglobin, 160
Metal, in food packaging, 253–255
Microbial cultures, 42–43, 134–135, 185
Microbial growth, 31–34, 42
    spoilage of food, prevention of, 222
    water activity, effect of, 36–37
Microbial fermentations. *See* Fermentation
Microbial life, forms of, 29–31
Microbial spores, 31
Micronutrients, 7–12, 22–23
    in processed foods, 23t, 246
    losses, 22
    requirements for, 10t, 18, 19t, 22–23
Microorganisms:
    detection and enumeration, 40–42
    fermentation processes, 42–45. *See also* Fermentation
    food-borne illness, 38–40
        outbreaks of, statistics, 290–291
    food poisoning, 38–40
    forms, 29–31
    in human nutrition, 44–45
    indicator, 39
    pathogenic, 38–39
    phage, 42
    psychrotrophs or psychrophiles, 33, 34
    requirements for growth, 32–34
    spores, 31
    starter cultures, 42–43
    temperature effects, 32–33
    total count, 40
Middlings. *See* Flour
Milk:
    antibiotics in, 129
    bulk tank collection, 125
    cattle breeds, 2
    clotting, by rennet, 135, 138
    composition, 8t, 127
    concentrated and dried, 146
    cream, separation of, 132
    cultured products, 133–137
    cheese products from. *See* Cheese
    dried, instantized, 147, 227–229
    fat content, in cream, 133t
    fluid, processing of, 127–133
    homogenization, 132
    HTST pasteurizer, 130–132
    human, composition of, 8t, 149
    lactose, 127, 281
        content, 8t, 128t
        digestion, 149
        in fermented products, 133
        intolerance, 149
        sandiness, in ice cream, 146
        solubility, 88t
    minor components, 127, 129t
    nutritive value of dairy products, 147–149
    pasteurization, 129–132
    payment to farmers, 127
    production, on farms, 125
    protein, 5, 127, 128t
        casein, 127, 128t, 134, 149
        casein micelles, 127
        components of, 5, 128t
        nutritive value, 18t
        sediment in UHT milk, 147
        in yogurt, 134
    in rat feeding experiments, 281–282
    as source of calcium, 148t, 149
    standardization of, 132
    sterilized, 147
    storage, silo tanks, 128–130
    UHT, 147. *See also* UHT processing
    vitamin content, 148
Milking parlor, 126
Milk powder, instantized, 227, 228
Milling, of wheat. *See* Flour, Wheat
Milling, of rice, 106
Minerals, 5, 19t
    analysis for, 6t
    stability in processing, 9
Modified atmosphere, for storage of foods, 76, 263–264. *See also* Fruits and vegetables, Food packaging
Moisture, in foods. *See* Water
Molasses, 88
Molds, 31, 37
    antimolding preservatives, 210t, 211
    in cheese-making, 139t
    spoilage, in bread, 104, 211
Mother cultures, in fermentation, 42–43
Muscle. *See* Meat
Mushrooms, 72t, 79t, 233

Mylar, plastics, 256t
Myoglobin, 160, 165
Myosin. *See* Meat protein

Niacin, 10t, 173
Nickel, in hydrogenation, 116
Nitrite, sodium, 165–166, 211, 294, 296, 297
Nitrogen:
    liquid, in freezing, 236
    in packaging, 192, 229, 263
Nitrosamines, 165
Nitroso-hemo-chrome, 165
Non-nutritive sweeteners, 194
Nutrients, 7–12
    essential, 13
    RDA, 19
Nutrition:
    disease, 13, 14t
    guidelines, 19
        Canada Food Guide, 20, 21
    labelling, 7, 214, 215t
    requirements, 13–19
Nutritive value:
    beverages, 195–197
    cereals and oilseeds, 119–121
    dairy products, 147–149
    dog food, 280–281
    fruits and vegetables, 88–90
    meat products, 172–173
    processed foods, 19–24, 246
    single foods, 277–286
    wine, 196
Nylon, in packaging, 256t

Oats, porridge of, 106
Obesity, 13, 120
Oil:
    composition, 111, 112
    content, in oilseeds, 110t
    crude, 113
    emulsions, 117–119
    extraction, techniques for, 113–114
    hydrogenation, 116
    vegetable, uses for, 114–116
Oilseeds, 93, 109–114
    Canola, 111
    composition, 8t, 113–114
    content of oil, 110t
    expeller pressing, 113–114
    nutritional importance, 119–121
    oil extraction techniques, 113–114
    protein, recovery from, 94, 114
    soybeans, 109, 110t
        protein, 114–115, 203
    storage requirements, 93
    rapeseed, 109
    world production, 110t
Orange juice, frozen concentrated, 83
Organic acids:
    in freezing, 233
    in fruits and vegetables, 82, 84
Organic foods. *See* Health foods
Osmophilic microorganisms, 33, 34, 37
Osteoporosis and osteomalacia, 14, 149
Overrun, in ice cream, 145
Oxalic acid, in fruits and vegetables, 89
Oxidation:
    antioxidant preservatives, 205t, 305
    in enzymatic browning, 81t
    of lipids, in dry foods, 229
Oxygen:
    in food spoilage, 221
    and meat color, 160, 262
    in microbial growth, 33
    respiration of fruits and vegetables, 73, 75–76

Packaging, food, 249–265. *See also* Food packaging, Packaging materials
Packaging materials, 250–259
    aluminum, 255
    cellophane, 256, 259
    flexible, for liquids, 257
    glass, 252–253
        bottles, returnable, 253, 263
    laminates, 259–262
        adhesion process, 259
        extrusion process, 259
        for UHT processing, 261
    metal, 253–255
    paper, 252–253
        glassine, 253t

Index 317

Packaging materials (*continued*)
    Kraft, 253t
        parchment, 253t
    plastics, 255–259
        blow-molding, 256
        extrusion, 256
        injection-molding, 256
        mil, definition, 259
        permeability, for gas, 257, 262
        printability, 257
        properties of, 257–258, 262
    sterilization of, 263
Pantothenic acid, 10t
Paper, in food packaging, 252, 253
Pasta products, 2, 105–106
Pasteur, Louis, 29, 236, 287
Pasteurization:
    of beer, 184
    of milk, 129–132
Pastry, baking, 105
Pathogenic microorganisms, 38–40
Pearson's square, 271–273
Pectin, 83t, 84
Pellagra, 14t
Penicillin, mode of action as drug, 211
Pennicilium camemberti, 139t
Penicilium roqueforti, 139t
PER, 18t
Peroxidase, enzyme, 80
Pesticides, 212t
pH:
    adjusting, food additives, 305
    in cheese-making, 136, 138
    of muscle, 156
    for pectin gels, 84
    of soft drinks, 194t, 195
Phage, 42
Phase diagram, 231
Phosphatase, enzyme, 130. *See also* Milk pasteurization
Phosphoric acid, in soft drinks, 195
Phosphorus, as micronutrient, 19t
Photosynthesis, 73
Pickling, 34, 243
Pimaricin, 210, 211
Plastics, in food packaging, 255–259
Polyunsaturated fats, 112
    in cardiovascular disease, 120–121, 144

Pond, aerobic, in waste treatment, 56t
Popcorn, 106
Pork:
    composition, 8t
    fat, saturation of, 112t, 172
    and Trichinosis, 162
Post-processing contamination, 132, 262
Potassium, as micronutrient, 19t
Potato chips, 82
Potatoes:
    browning of, 81
    French fried, 2, 3, 82
    processing of, 76–77
    reconditioning, 77
    suberisation, 76
    suitability for processing, 2
Poultry:
    breeding, 3
    fat, saturation of, 172
    meat, composition of, 8t
    mechanical separation ("deboning"), 170
    processing of, 169–170
Preservation of food. *See* Food preservation
Preservatives, as food additives, 305. *See also* Food additives
Printability, of packaging materials, 257
Processed cheese, 141
Processed foods. *See specific entries*
Product development, in food industry, 59
Proof, degree of, 189
Proofing, of bread, 101
Propionic acid, 211
Protein:
    analysis for, 6
    changes in cheese ripening, 140
    composition, 14, 16
    concentrate, 114
    energy content, 15t
    isolate, 114
    meat, 155t
    milk, 127–128
    nutritive value, PER, 18t
    oilseeds, 114
        amino acid deficiency of, 121
    PER, 18t
    single-cell, 44–45
    soy, 114–115, 203

Protista. *See* Microorganisms
Proximate composition, 4–6
    gravimetry, 5
    orientation values, for foods, 8t, 9t
Psychrophiles or psychrotrophs, 33, 34

Quality, characteristics of:
    dried foods, 224–225
    french fries, 82
    frozen foods, 230
    heat-processed foods, 240
Quality control, 59–62
    examples of procedures, 62t, 63t
Quarg, 135

Radappertization, 243
Radiation, for food preservation. *See* Food irradiation
Radiation sterilization:
    medical disposables, 244, 263
    packaging materials, 263
Radurization, 243
Rancidity, 97, 223, 228
Raoult's Law, 36
Rapeseed, 109. *See also* Oilseeds
Rausing, Ruben, 240
RDA, 19
Red No. 2 and No. 40, 207, 208
Refractive index, of sugar, 85, 86t
Refining, of sugar, 88
Refractometer, 84–85
Refrigeration:
    in food preservation, 241–242
    mechanical, system, 235
Rehydration, of dried foods, 225–227
Relative humidity:
    in fruits and vegetables storage, 72
    in meat aging, 157
Rennet or rennin, clotting of milk, 135, 138
Respiration, in fruits and vegetables, 73, 75–76
    anaerobic, 75
Retort, 238–239
Retortable pouch, 240
Retrogradation of starch, in bread, 104
Riboflavin (Vitamin B-2), 10t, 24, 133, 148
Rice, milling of, 106

Rickets, 14t
Rickettsia, 38
Rigor mortis. *See* Meat
Risk-benefit:
    of food additives, evaluation, 208, 210t
    of meat curing, 166
Roquefort cheese, 137t, 139t. *See also* Cheese
Ruminants, efficiency of food production, 153–154
Rye:
    crossed with wheat, triticale, 106
    flour in bread, 103
    for liquor manufacture, 177t
    milling, 106

Saccharin, 208
Sacharomyces yeast, 177, 184t, 185
Safety, of food. *See* Food
Salmonella:
    characteristics of illness, 39t
    in food-borne disease outbreaks, 290t
    in poultry, 162, 170
    sources of, 38
Salt:
    in cheese-making, 139–140
    as humectant, 217
    in meat products, 167t, 298
    and water activity, 35
Sandiness, in ice cream, 146
Sanitary can:
    production of, 238–239
    identification, 255t
Saran, plastics, 256t
Saturated fats, 111, 112t
Sauerkraut, 42, 243
Sausage. *See* Meat
Scurvy, 14t
Seafood, sources of, 171t
Sensory evaluation, 59–61, 63
Shellfish, in food poisoning, 293
Sherbet, 145t
Shigella, in food-borne illness, 38
Shortening, 114, 117
Single cell protein (SCP), 44–45
Skim milk, 53, 67t, 132, 226
Smoking, of meat, 169

Sodium, as micronutrient, 19t
Sodium benzoate, in soft drinks, 195
Sodium bisulfite, and browning potatoes, 81
Sodium nitrite, 165–166. *See also* Meat
Soft drinks, 192–195
    acidity, 195
    aspartame in, 194
    caffeine content, 194
    carbonation of, 195
    composition of, 194
    consumption of, 192
    diet, 193–194
    fructose in, 194
    pH of, 194t
Solanine, 206, 215
Solubility, 68
    of sugars, 88t
Sorbic acid, 211
Sour-dough, bread, 101
Sour cream, composition of, 133t
Sour milk. *See* Milk, cultured products
Soybeans, 109, 114. *See also* Oilseeds
    protein use, 114, 203
Specific heat, 274
Spinach, in rat-feeding experiments, 283
Spoilage of food. *See* Food spoilage
Sponge dough, 101
Spores, microbial, 31. *See also* Microorganisms
Spray drying, 225–227
Stabilizers, in ice cream, 45, 145–146, 202
Staphylococcus aureus, 38, 40
    in food-borne disease outbreaks, 290t
Starch:
    brewing, breakdown in, 178
    gelatinization, by heat, 102, 104
    modifying agents, as food additives, 306
    retrogradation in bread, 104
Starter cultures, cheese, 138
Steel, for food cans, 254
Sterility, commercial, 237
Sterilization:
    of foods, 222t, 236–246
    of medical supplies, by irradiation, 244, 263
    of milk, 147
    of packaging materials, 263

Sterilizer, hydrostatic, 239t
Stokes' Law, 118
Streptococcus thermophilus, 134, 135t
Sublimation, in freeze-drying, 225
Sucrose. *See* Sugar
Sulphur dioxide, and browning of dry fruits, 81
Sugar:
    crystallization of, 85, 86–88
    as a food additive, 215
    as humectant, 217
    in jams and jellies, 83–85
    in soft drinks, 193–194
    in wine, 187–188
Sun drying, 223, 224t
Surfactants, 118, 119
Surimi, 171
Sweeteners:
    aspartame, 194
    fructose, in soft drinks, 194
    non-nutritive, as additives, 304
    saccharin, 208
Swiss cheese, 137t, 139t, 141. *See also* Cheese

Tallow, saturation of, 112t
Tannins, in tea, 190
Taste panel, 59
Tea, 189–192
    caffeine content, 192
    fermentation of, 190
    flavor compounds in, 191
    instant, 191
    oolong, 190
Temperature:
    and microbial growth, 32–33
    of milk pasteurization, 130
Tenderizers, meat, 157
Tetra-pak, 240
Texture:
    of bread, 102, 103
    in food freezing, 230, 233
    of ice cream, 145
    modifying, food additives, 304
Thiamin, 10t, 24, 173
Thickening agents, 45, 146, 205t
Tin, in can manufacture, 238

Tomatoes:
    mechanical harvesting, 3
    pre-processing for ketchup manufacture, 77
Trichinella spiralis, 162
    in food-borne disease outbreaks, 290t
Triglycerides, 111
Triticale, 106

UHT processing, 147, 239–241
    direct heating, 240
    hydrogen peroxide, for package sterilization, 240
    packaging materials, 261
    post-processing contamination, 262
    Tetra-pak, 240
Unsaturated fats, 111, 112, 120–121
USDA, 64t, 65

Vacuum:
    in cans, 236
    drying, 224t
    oven for moisture content analysis, 6t
Vapor pressure. *See* Water
Vegetable oils, composition of, 111–112. *See also* Oils
Vegetables, botanical characteristics, 71–72. *See also* Fruits and vegetables
Vinegar, 243
Vitamin A, toxicity of, 10t, 215
Vitamin B group, losses in milling, 96–97
Vitamin C:
    freezing, losses in, 236
    in fruits and vegetables, 22, 88–89
    in milk, 24, 148
    potatoes, French fried, 82
    processing losses, 22, 24, 89, 148
    in rat-feeding experiments, 282
Vitamin D:
    and calcium absorption, 12
    milk fortification, 148
Vitamins:
    biological activity, 9
    enrichment of flours, 97
    in heat-processed foods, 240–241
    in meats, 173
    as micronutrients, 9, 12
    in milk, 148
    in processed foods, 12, 246
    in proximate analysis, 4
    requirements for, 10–11t
    sources of, 10–11t
    stability 9–11
    supplement pills, 12

Waste treatment, 53–54
Water:
    analysis for, in foods, 6
    in beer-making, 178
    content in foods, orientation values, 8t
    in food processing, 53
    free, 35
    hygroscopic foods, 229
    reabsorption, by dry foods, 225
    vapor pressure, 35–36
Water activity, 34–37
    control, for food preservation, 243
    definition, 35
    humectants, 37, 216–217
    intermediate moisture foods, 37, 216
    and microbial spoilage, 36–37
    Raoult's Law, 36
Water soluble vitamins, 10–11t
Wheat, 94–99
    bran, 94–96, 121
    durum, 94, 105
    flour, gluten content, 101
    germ, 97
    kernel, structure of, 94
    milling technology, 98–99
    whole, flour, 97, 99
Whey:
    from cheese-making, 135
    drinks, 149, 150t
    protein in yogurt, 134
Whippability, 202–204. *See also* Functional ingredients
Whipping agents, as food additives, 306
Wine, 185–188
    aging, 187
    Baby Duck, 187
    fermentation of, 185–187
    fortified, 188

Wine (*continued*)
    nutritional aspects, 196
    world production, 186
    Zinfandel, 187
Wort, in beer-making, 181–183

Yeasts:
    in bread-making, 100–102
    in brewing, 183–184
    morphology, 30–31
    in wine-making, 185–186

Yeast foods, as food additives, 306
Yersinia enterocolitica, 39
Yogurt:
    nutritive value, 148–149
    technology, 134
    whey protein in, 134

Zinc, as micronutrient, 19t
Zinfandel, 187